Learning Basic Math and Business Math Using the Calculator

2d Edition

Barbara F. Muncaster
Accounting Professor
Rose State College
Midwest City, Oklahoma

Susan Prescott
Adult Education Instructor
Frances Tuttle Vocational
Technical School
Oklahoma City, Oklahoma

ISBN: 0-538-60815-3

Library of Congress Catalog Card Number: 90-61648

8 9 0 H 0 9

Printed in the United States of America

Acquisitions Editor: Bob Lewis
Developmental Editor: Willis Vincent
Production Editor: Angela Crum
Designer: Darren Wright
Production Artist: Nicole Jones
Photo Researcher: Michale O'Donald
Marketing Manager: Greg Getter

Preface

Welcome to the world of math. Almost every job in today's business world requires mathematical skill. LEARNING BASIC MATH AND BUSINESS MATH USING THE CALCULATOR will help you acquire this skill. Employers often complain that some people do not understand basic mathematical processes and how they are applied to business problems. Employers generally do not pay employees to simply add, subtract, multiply, or divide numbers. They want employees who know how to apply basic math to business situations. A calculator can aid in performing mathematical operations by insuring accuracy in many computations. However, you need to understand the math operations, or a calculator is of little value.

LEARNING BASIC MATH AND BUSINESS MATH USING THE CALCULATOR is a text-workbook designed to develop the math skills necessary for solving business problems using a calculator. This book is organized into 18 units and 4 retail applications and requires one semester to complete. Class instruction may be on a group basis or an individual basis. A teacher's edition, a pretest, ten unit tests, and a posttest accompany the text.

Units 1 through 4 cover the basic mathematical operations: addition, subtraction, multiplication, and division. Units 5 through 8 provide other basic math concepts dealing with fractions, decimal fractions, multioperations, decimals, percents, and percentages. Units 9 through 13 provide math concepts related to discounts, banking, and payroll. Units 14 through 18 discuss the math involved in pricing, depreciation, credit, ratio and proportion, and investments.

Each group of units is followed by a retail simulation. These four retail applications provide an opportunity to apply the concepts presented in the previous units to business situations.

The increase in international trade has created a need for people to better understand metric measurement. Problems have been added to this edition that require knowledge of metric measurement and use of a conversion table. Word problems given at the end of most units help develop an awareness of the international market.

Each unit in LEARNING BASIC MATH AND BUSINESS MATH USING THE CALCULATOR is organized in the following way:

Objectives: Each unit begins with the objectives to be learned.

Math Terms and Concepts: The fundamental math concepts are presented with a list of terms and definitions, a brief explanation of the operations, and drill practice for reinforcement. Reminders and hints covering concepts are indicated with an asterisk or placed in the margin for emphasis.

Calculator Instructions: Step-by-step instructions are explained for calculator operations. Exercises are given to apply the calculator operations in Units 1 through 6. Correct answers are included for some problems so results can be checked immediately.

Evaluating Your Skills: Units 1 through 4 evaluate the use of the calculator while Units 5 through 18 have an accumulative review of all concepts presented.

Applying Your Knowledge: Exercises are given that apply the math principles in each unit to realistic business situations, forms, and reports.

Advanced Applications: More advanced knowledge of business math concepts must be applied in these exercises.

The appendices can be used for general calculator instructions and for help with metric conversions, decimal equivalents, and multiplication using a table.

A special thank you to Judy O'Hare, owner of Sweet Ideas, Oklahoma City, Oklahoma, for contributing current information used in the retail applications.

Barbara Muncaster
Susan Prescott

Contents

unit 1

Addition of Whole Numbers and Decimals

"Time is a treasure— use it wisely."

Addition is used in almost all business calculations. Understanding the operation is vital in solving daily business problems. Once you understand the addition operation, you can use the calculator to save time.

Using the touch method on the calculator will further decrease the time needed to complete business problems. Often, businesses require applicants for jobs to pass a timed test using the touch method on the electronic calculator or computer 10-key pad.

After studying this unit, you should be able to:

1. identify the place value of numbers
2. add whole numbers and decimals and verify sums
3. operate the calculator using the touch method
4. use the calculator to add whole numbers and decimals
5. solve business math problems using addition

MATH TERMS AND CONCEPTS

The following terminology is used in this unit. Become familiar with the meaning of each term so that you understand its usage as you develop math skills.

1. **Addend**—the number(s) to be added to another.
2. **Addition**—the process of combining two or more numbers.
3. **Bank Statement**—a periodic report from a bank to anyone having an account that lists all checks and deposits processed, bank charges, and the balance.
4. **Decimal**—any number to the right of the decimal point.
5. **Decimal Point**—a period separating a whole number and a number less than one.
6. **Digit**—a single number (1, 2, 3, 4, 5, 6, 7, 8, 9, or 0).
7. **Number**—a digit or group of digits.
8. **Number Sentence**—a statement that says two numbers are equal by the use of the equal sign (=).
9. **Numeral**—a figure, symbol, or name used to express a number.
10. **Place Value**—the value of a digit determined by its place within a number.
11. **Sum**—the number obtained by adding two or more addends—the answer; the total.
12. **Verify**—check the accuracy of data.
13. **Whole Number**—any number to the left of the decimal point.

The value of a digit is determined by its *place* within a group of digits. Each place is ten times greater than the place to its right. The following **place value** chart identifies the name of each place to the left of the decimal point. Notice the number of places between each digit and the decimal point.

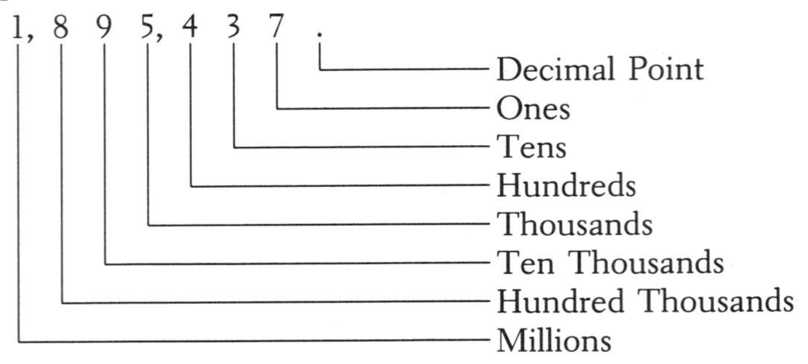

The comma is used for ease in reading numbers. Beginning at the right, a comma is placed after every third digit to the left of the decimal point.

On the lines below, write the digit in the number 954,183 that has the place value indicated.

__________ ones
__________ tens
__________ hundreds
__________ thousands
__________ ten thousands
__________ hundred thousands

The decimal point is understood in a **whole number**; therefore, the decimal point does not have to be written. The number 426 could be written 426. or 426.0. Because the zero to the right of the decimal does not have any value, we can drop the zero and the decimal point.

The **numerals** 1, 2, 3, 4, 5, 6, 7, 8, and 9 positioned to the right of the decimal point indicate a part of the whole. They are referred to as **decimals.** Each digit to the right of the decimal point has a place value in relation to its portion of the whole. The name of each place to the right of the decimal point is shown in this example:

. 2 7 9 1 2
Decimal Point
Tenths
Hundredths
Thousandths
Ten Thousandths
Hundred Thousandths

On the lines below, write the digit in the number 183.457 that has the place value indicated.

__________ thousandths
__________ hundredths
__________ tenths
__________ ones
__________ tens
__________ hundreds

Addition is the process of combining two or more numbers (**addends**) to find a total (**sum**).

Example:
$$\begin{array}{rl} 56 & \text{addend} \\ +87 & \underline{\text{addend}} \\ \hline 143 & \text{sum} \end{array}$$

The order in which numbers are added does not affect the sum.

Examples:
$$\begin{array}{r} 7 \\ +9 \\ \hline 16 \end{array} \qquad \begin{array}{r} 9 \\ +7 \\ \hline 16 \end{array} \qquad \begin{array}{l} 6+3+2=11 \\ 2+6+3=11 \\ 3+2+6=11 \end{array}$$

Reversing the order in which numbers are added will **verify** addition.

Example: Add / Verify (Add up)
$$\begin{array}{r} 4 \\ +2 \\ +6 \\ \hline 12 \end{array} \qquad \begin{array}{r} 4 \\ +2 \\ +6 \\ \hline 12 \end{array} \uparrow$$

Adding zero (0) to any number does not change the value of the number.

Examples:
$$\begin{array}{r} 0 \\ +8 \\ \hline 8 \end{array} \qquad \begin{array}{r} 7 \\ +0 \\ \hline 7 \end{array}$$

In the following exercises, fill in the blank with a number that will complete the **number sentence**.

1. 3 + __________ = 3
2. 25 + 96 = 96 + __________
3. __________ = 44 + 0
4. If 7 + 6 + 2 = 15, then 2 + 6 + 7 = __________
5. If 16 + 10 + 23 = 49, then 23 + 16 + 10 = __________
6. 101 + 32 + 764 + 1,589 = 764 + 32 + __________ + 101

Digits of the same place value are aligned vertically for ease in adding. Decimal points are always aligned. Look at examples a, b, c, and d. Which problems could you add quickly without errors?

a.	b.	c.	d.
375	375	375.95	375.95
450	450	450.36	450.36
21	21	.21	.21
4,098	4,098	4,098	4,098.00

If you answered *b* and *d*, you are correct because the whole numbers and decimals are aligned. * ***Note that in example d, zeros were added to the right of the decimal point without changing the value of the number (4,098 = 4,098.00). This is done for ease in alignment.***

Decimals in the sum are placed directly below the decimals in the addends.

Example:
$$\begin{array}{r} 355.55 \\ +\quad .21 \\ \hline 355.76 \end{array}$$

Carrying

When adding numbers containing two or more digits, it may be necessary to carry a number to the next column.

Example:

	Add ones column	Add tens column	Verify (Add up)
33	3	1 33	1 33 ↑
+59	+9	+59	+59
	12 (Carry 1 from the ones column to the tens column.)	92	92

Learn to carry mentally!

Properly align the addends for addition, add zeros where needed, and find the sum.

1.
```
  42.59
    861
  76.17
  143.6
+    .6
    sum
```

2.
```
   .388
  1,406
    .06
+  78.9
    sum
```

Vertical and Horizontal Addition

In each of the following problems, arrange the numbers vertically, add zeros where needed, and find the sum.

1. 7.21 + 85 + 3.1 + .916

2. 72.17 + 2.509 + 42 + 1,100

3. 40.52 + 665.36 + 3,712 + 5.08

4. .007 + 1.6 + 23 + 2

Are your numbers written clearly? Mistakes are often made when numbers are written carelessly!

Practice

Find the sum in each of the following problems. Work the problems as quickly as you can.

Do not use your calculator!

1.
```
  6
  5
 11
```

2.
```
 4
 3
```

3.
```
 3
 4
```

4.
```
 45
 63
```

5.
```
 28
 37
```

6.
```
 89
 43
```

7.
```
 78
 63
```

8.
```
 8,127
 9,052
   367
    29
```

9.
```
   125
 2,763
   875
```

10.
```
 414
 372
 905
 681
```

11. 35.76 + 1.24 + 16.92 + 4.18 = _____ **12.** 72.1 + .61 + 24.36 = _____

Did you verify your answers?

13. In the number 21.98 the digit in the ones position is ___________ and the digit in the hundredths position is ___________.

CALCULATOR INSTRUCTIONS

"The calculator is only as smart as the operator."

* ***Turn to Appendix A, page 335. Read the general calculator instructions. Refer to your calculator manual for operating instructions before asking your instructor.***

Locate the following on your calculator:

On/Off Switch

Paper Advance

Decimal Selector

Add Mode Position

Total Key

Plus Bar or Key

Clear Entry Key

Minus Key

Touch Method

To operate the calculator by *touch,* use the first three fingers and thumb of the right hand. (Some people can operate a calculator with their left hand while writing with their right hand. Try it! If you are comfortable, use your left hand to operate your calculator. Appendix A, page 335, explains the touch method for using the left hand.)

Solve the problem: 4 + 5 + 6 =

1. Turn on your calculator.
2. Set the decimal selector on 0. (If your calculator does not have a 0 position, set the decimal selector on 2 and disregard the printed decimal.)
3. Check your posture!
4. Clear your calculator. Operate the total key once or the clear key twice.

   * ***Develop a good habit of always looking for the total symbol on the printout tape or a clear indicator on the screen before beginning any problem.***
5. Curve your index, middle, and third fingers of your right hand and place them over the 4, 5, and 6 keys. This is your *home row.*

6. Enter the 4 with your index finger.
7. Enter the addend into the machine by operating the plus key with your little finger.
8. Enter the 5 with your middle finger and add by operating the plus key.
9. Enter the 6 with your third finger and add.
10. Obtain a total by operating the total key once.
    Did you get 15 as the sum?

* ***Check your calculator manual for specific operating instructions.***

Pocket Calculator

Due to its size, the pocket calculator is usually not operated by the *touch method.* However, you should follow the operating instructions in the calculator manual, practice the problems, and take the tests to develop proficiency in using your pocket calculator.

Practice

Keep your eyes on the book as you find the sum of the following problems using the touch method. Keep one hand free for following figures. Using a stiff piece of cardboard, move down the columns as you add the numbers. This will assist you in keeping your place.

Home Row: 4, 5, and 6 Keys

1.	2.	3.	4.	5.
4	45	564	56	66
56	4	44	664	54
5	65	46	65	55
65				

6.	7.	8.	9.	10.
546	464	644	556	454
64	55	564	466	645
54	565	465	54	456
46	45	64	44	4,564
564	56	56	645	645
55	64	55	545	554
645	646	546	664	66

To clear a wrong entry before the plus key has been used, operate your *clear entry key* once. If you have already operated the plus key, operate your minus key immediately. It is not necessary to reenter the wrong numbers before depressing the minus key.

cathy® **by Cathy Guisewite**

Extension: 1 and 7 Keys

Use your index finger to reach from the 4 key up to the 7 key or down to the 1 key. The next 10 problems give practice using the 1, 4, 5, 6, and 7 keys. Remember to keep your fingers curved and your eyes on this book!

1.	2.	3.	4.	5.
4	77	14	74	57
47	71	11	447	15
41	4	177	741	571
92				

6.	7.	8.	9.	10.
61	156	45	156	166
156	766	657	716	5,741
745	1,476	717	6,741	6,611

Did you keep your eyes on the book? If not, repeat the exercise above.

Extension: 2 and 8 Keys

Use your middle finger to reach from the 5 key up to the 8 key or down to the 2 key. The next 10 problems give practice using the 1, 2, 4, 5, 6, 7, and 8 keys.

Keep your eyes on the book!

1.	2.	3.	4.	5.
82	258	28	22	821
25	28	25	822	584
28	255	88	252	52
85	82	828	85	18
220				

6.	7.	8.	9.	10.
488	224	627	668	86
87	52	75	862	568
718	174	821	456	678
77	852	782	28	25

Extension: 3, 9, and 0 Keys

Use your third finger to reach from the 6 key up to the 9 key or down to the 3 key. Use your right thumb to operate the 0. The next 10 problems give practice using all the keys. * ***If your calculator has a double 0 key (00), it may be easier to operate it with your middle finger rather than your thumb.***

1.	2.	3.	4.	5.
796	960	33	30	895
93	992	903	90	693
639	639	660	603	253
1,528				

6.	7.	8.	9.	10.
343	7,940	3,009	49	360
840	64	500	99	650
653	34	4,539	460	934
4,895	1,503	180	7,646	1,327

Did you use the touch method?

Always *verify* your answers.

Printing Calculators: Compare the tape entries with the problem or use reverse addition.

Computer Calculators: Print the tape and compare with the problem or use reverse addition.

Display Calculators: Repeat the problem by using reverse addition.

Add each of the following problems. When you obtain the correct total using the touch method, go on to the next problem. Repeat a problem until your total is the same as the printed total.

11.		12.		13.		14.		15.	
	5,270		7,068		163		7,549		1,762
	2,739		1,736		2,363		8,460		5,985
	5,895		245		4,887		8,090		9,372
	267		9,867		5,799		7,007		2,570
	14,171		18,916		13,212		31,106		19,689

All Keys

Remember to verify your answers.

1.		2.		3.		4.		5.	
	90		82		390		163		36
	96		25		769		58		14
	267		49		13		41		708
	51		375		81		25		75
	60		892		40		95		37

6.		7.		8.		9.		10.	
	85		198		28		840		651
	43		30		898		3		388
	93		445		206		600		907
	374		60		582		971		724
	31		274		434		494		25

11.		12.		13.		14.		15.	
	6,851		8,232		9,340		7,642		6,160
	191		2,057		1,476		4,325		78
	7,155		9,231		908		9,151		2,718
	357		949		890		4,658		2,723
	6,784		4,613		6,118		3,965		5,300

Addition of Varying Decimals

To add numbers with the same number of decimal places, set the decimal selector on the number of places needed. If your decimal selector does not have the exact places required, set it on the next highest setting. Use your third finger to operate the decimal point.

Examples:

4.1	(Set the	3.77	(Set the	3.899	(Set the
2.4	decimal	4.65	decimal	1.283	decimal
+5.9	selector	+8.05	selector	+6.304	selector
	on 1.)		on 2.)		on 3.)

To add numbers with varying decimals, set the decimal selector to accommodate the number with the most decimal places.

Remember: Decimal refers to all numbers to the right of the decimal point.

Solve the problem:

```
   23.1
  202
+  14.235
```

1. Set the decimal selector on 3. * *Use 4 if your calculator does not have 3.*
2. Check your posture.
3. Clear the calculator.
4. Enter 23.1 and operate the plus key.
5. Enter 202 and operate the plus key. * ***It is not necessary to enter the decimal point and zeros when adding whole numbers.***

6. Enter 14.235 and operate the plus key.
7. Obtain a total by operating the total key once.
   Did you get 239.335 as the sum?

* ***Notice the calculator has added zeros for proper alignment.***

Find the sum for each of the following problems.

* ***In each problem, set the decimal selector to accommodate the number with the most decimals. If your calculator does not have a setting for the exact number of places required, set the selector to a higher setting. If more decimals are required than your decimal selector has, set the decimal selector on floating decimal (FL). The decimals will not be aligned in this setting.***

Remember to enter the decimal point.

1.	**2.**	**3.**	**4.**	**5.**
105.96	23.48	432.17	597.32	71.6987
28.561	5.43	86.73	21.725	156.7434
13.78	15.24	1,140.5	657.017	4,902.08
276.47	321.652	684.1	54.42	395.372
82.55	654.2347	27.14	567.98	2,199.8
29.46	894.618	357.89	361.59	57.49
3,037	40.992	100	11.52	668.567
183.9	116.7	74.221	152.668	48.913
4,349.27	56.46	159.973	860	339.56
541.016	540.48	76.4	88.968	88.357

Addition of Dollars and Cents

Some calculators have an *add mode* position on the decimal selector while other calculators have a separate add mode selector. The add mode can be set to add numbers with two decimal places (dollars and cents). By setting the calculator's add mode, and decimal selector if required, you will not have to enter the decimal point in each number. The decimal point will automatically be placed between the dollars and cents. * ***If you do not have the add mode position, set the decimal selector on 2 and use your third finger to operate the decimal point.***

Find the sum for each of the following problems.

When entering whole numbers, it is necessary to enter double zeros for the decimal places.

1.	**2.**	**3.**	**4.**	**5.**
$ 25.72	$ 70.09	$ 90.30	$ 25.70	$ 58.88
39.99	80.80	4.00	58.50	25.62
4.61	6.30	83.00	63.90	8.24
46.59	7.04	67.78	.43	31.94
$116.91				

6.	**7.**	**8.**	**9.**	**10.**
$ 2.63	$ 1.65	$ 24.46	$ 346.40	$3,822.00
2.00	87.00	59.70	623.42	4,039.65
9.46	67.37	19.21	53.87	125.98
1.63	91.15	775.11	819.15	637.17

Are your answers legible?

* *Did you write a dollar sign in your answers?*

Addition with Repeated Numbers

Most calculators have a *repeat entry* feature. Once a number is entered into the calculator it can be repeated by simply operating the plus key. The number does not have to be entered a second time.

Solve the problem:

```
  25
  30
  30
+ 73
```

1. Set the decimal selector on 0.
2. Check your posture.
3. Clear the calculator.
4. Enter 25 and operate the plus key.
5. Enter 30 and operate the plus key twice.
6. Enter 73 and operate the plus key.
7. Obtain a total by operating the total key once.
   Did you get 158 as the sum?

The problem in the example does not have any decimals; therefore, the decimal selector is set on 0. Look at each problem to be added and then decide where to set the decimal selector.

Add each of the following repeated-entry problems. * ***Notice the problems have two decimal places. Set your calculator's add mode or set the decimal selector on 2.***

1.	**2.**	**3.**	**4.**	**5.**
4.28	632.27	564.70	141.79	62,011.00
1.28	184.37	851.02	141.79	62,011.00
1.28	184.37	56.55	72.45	62,011.00
4.23	184.37	757.49	965.30	56,495.75
5.13	211.15	86.66	29.72	56,495.75
5.13	211.15	86.66	29.72	9,571.60
5.13	420.25	955.73	779.47	4,892.20
5.22	17.53	955.73	59.45	13,476.98
3.26	66.40	10.80	59.45	13,476.98
3.26	66.40	337.73	808.00	2,891.46

Vertical and Horizontal Addition

In vertical and horizontal addition, add the columns vertically (↕) and horizontally (↔). Then find the sum of the vertical column totals and the sum of the horizontal column totals. The two totals should equal. If they do equal, you have verified the accuracy of your addition. * ***Set decimal selector on 0.***

1.

823	259	379	724	209 =	_____	**a.**
396	968	911	949	392 =	_____	**b.**
603	180	517	954	523 =	_____	**c.**
537	875	178	754	536 =	_____	**d.**
f. _____	**g.** _____	**h.** _____	**i.** _____	**j.** _____	_____	**e.**

(vertical column total)

_____ **k.**
(horizontal column total)

2.

671	704	423	589	823 =	_____	**a.**
565	808	923	500	500 =	_____	**b.**
279	603	281	727	258 =	_____	**c.**
e. _____	**f.** _____	**g.** _____	**h.** _____	**i.** _____	_____	**d.**

(vertical column total)

_____ **j.**
(horizontal column total)

EVALUATING YOUR SKILLS

You will be timed on the following problems. Your minimum goal is to complete the problems in 10 minutes. Work each problem once.

1.	2.	3.	4.	5.
48.58	97.13	13.63	41.80	52.91
71.56	25.87	90.50	45.33	48.62
25.63	72.41	20.91	90.32	85.79
75.86	74.86	56.50	66.56	59.14
67.24	40.54	20.57	21.29	86.12
34.48	40.39	50.26	42.70	33.15
50.20	86.81	68.26	52.60	42.87
76.42	50.41	58.01	40.13	54.82
44.54	33.42	12.65	87.33	44.23
78.37	16.43	51.73	63.51	96.93

6.	7.	8.	9.	10.
69.66	84.64	73.28	61.03	92.57
38.86	25.05	21.13	71.38	65.00
97.37	39.40	60.73	70.72	50.22
62.00	29.64	28.11	33.64	52.58
53.67	95.39	36.47	97.14	91.33
15.65	42.96	22.50	46.21	10.26
51.99	48.90	23.11	10.73	72.10
36.27	56.34	95.53	92.15	42.80
41.75	96.84	28.67	19.52	78.43
76.65	89.78	32.35	24.61	57.00

If you were unable to reach your goal, or if you want to set your goal higher, practice the problems on page 10 to develop speed and accuracy. Then repeat the problems above.

Grading Scale

Completion Time	0 Errors	1 Error	2 Errors	3 Errors
5½ minutes	A+	A	B+	B
6 minutes	A	B+	B	C+
6½ minutes	B+	B	C+	C
7 minutes	B	C+	C	C−
7½ minutes	C+	C	C−	D+
8 minutes	C	C−	D+	D
8½ minutes	C−	D+	D	D−
9 minutes	D+	D	D−	repeat
9½ minutes	D	D−	repeat	
10 minutes	D−	repeat		
10½ minutes	repeat			

Grading Scale—End of 5 Weeks

Completion Time	0 Errors	1 Error	2 Errors	3 Errors
4½ minutes	A+	A	B+	B
5 minutes	A	B+	B	C+
5½ minutes	B+	B	C+	C
6 minutes	B	C+	C	D+
6½ minutes	C+	C	D+	D
7 minutes	C	D+	D	D−
7½ minutes	D+	D	D−	repeat
8 minutes	D	D−	repeat	
8½ minutes	D−	repeat		
9 minutes	repeat			

Grading Scale—End of 9 weeks

Completion Time	0 Errors	1 Error	2 Errors	3 Errors
3½ minutes	A+	A	B+	repeat
4 minutes	A	B+	B	repeat
4½ minutes	B+	B	C+	repeat
5 minutes	B	C+	C	repeat
5½ minutes	C+	C	C−	repeat
6 minutes	C	C−	D	repeat
6½ minutes	Practice, then repeat test!			

APPLYING YOUR KNOWLEDGE

When working word problems, read each problem carefully. Think about what you are trying to determine. Then read the problem again noting each figure you must use to solve for the answer.

Complete the following problems using your calculator.

1. The number of Cable TV subscribers in five cities was 54,310; 27,813; 19,416; 35,688; and 1,500. What was the total number of subscribers in the five cities?

____

2. Computer's Etc. quoted Janet the following prices on a home computer: PC with 64K and one disk drive—$1,600; monochrome monitor—$199, or color monitor—$549; second disk dive—$439; software package—$250. Janet decided to purchase the PC computer with color monitor and second disk drive. What is the total purchase price?

____

3. Carmelo needs to purchase various colors of ribbon to use in a display window. The display calls for 5.6667 yards of white, 3.6 yards of pink, 2.3333 yards of blue, 2.25 yards of yellow, and 4.125 yards of navy. What is the total number of yards needed?

____

4. In 1989 Mexico purchased 403,000 metric tons of U.S. corn and China bought 150,000 metric tons of corn and 100,000 metric tons of wheat. This was the first sale of corn to China in two years. What was the total number of metric tons of corn sold to the two countries? * *A metric ton is a measure of weight equal to 2,204.62 pounds.*

____

5. Phong, an accounting clerk for a sewing machine manufacturer, prepares a production report of the number of individual items produced each week. Add the columns vertically to determine the number of items produced for each week in April. Add the rows horizontally to determine the number of each item produced for the month. Then add the item totals to obtain a grand total. Verify the grand total by adding the weekly totals.

Production Report for April

Item	Week 1	Week 2	Week 3	Week 4	Totals
Bobbins	776	629	598	810	______
Needles	1,055	1,234	960	987	______
Seam Guides	598	622	427	630	______
Needle Plates	430	528	516	497	______
Stitch Feet	290	384	471	356	______
Weekly Totals	______	______	______	______	______

6. Find the following salesperson's first quarter sales and record the subtotal. Then add in the second quarter sales and record the total sales earned for the first half of the year. Also, find the monthly totals and the grand total for the first half of the year.

	Jan	Feb	March	First Quarter Subtotal	April	May	June	First Half Total
Ribbons	240	237	265	**a.** ______	266	314	280	**b.** ______
Cables	349	368	297	**c.** ______	385	320	376	**d.** ______
Disks	325	516	498	**e.** ______	471	384	412	**f.** ______
Manuals	117	256	193	**g.** ______	290	316	185	**h.** ______
	i. ______	**j.** ______	**k.** ______		**l.** ______	**m.** ______	**n.** ______	**o.** ______

7. In one year, U.S. domestic exports to various countries were as listed below. Find the total dollar amount exported and the total amount exported to each country by adding vertically and horizontally.

U.S. Domestic Exports

In Millions of Dollars

Commodity	Canada	Mexico	United Kingdom	Germany	Japan	Totals
Food & Live Animals	1,550	668	436	586	4,691	______
Beverages & Tobacco	58	5	60	207	865	______
Crude Materials	1,672	1,037	656	1,101	4,954	______
Mineral Fuels	1,360	510	154	63	297	______
Manufactured Goods	4,621	1,509	897	1,769	296	______
Machinery	13,267	5,372	5,558	4,117	4,794	______
Transport Equipment	15,841	1,557	1,963	1,375	2,283	______
Totals	______	______	______	______	______	______

ADVANCED APPLICATIONS

Calculate the answers for the following problems.

1. Tracy Wiley had to meet three clients on Tuesday. He drove a distance of 47 miles from his office to meet the first client. The second client was 24 miles further, and the third client another 30 miles from his office. He then drove back to his office.

a. How many miles did he drive? __________

b. If 1 mile equals 1.6 kilometers, about how many kilometers did he drive? * *Use the conversion chart in Appendix B, page 339.* __________

2. Chong Consulting checks its bank statement each month to find the total amount of deposits made and the total amount of checks written and processed by the bank. Deposits are indicated on the statement by the abbreviation *dep.* All other amounts are *checks*.

Date	Amount	Date	Amount
3/18	349.49-dep	4/07	63.00
3/23	18.75	4/08	19.52
3/23	47.56	4/09	16.00
4/01	20.00	4/10	40.00
4/01	1,036.63-dep	4/13	42.00
4/03	177.00	4/14	55.93
4/05	13.25	4/19	150.00-dep
4/06	95.00	4/20	79.42
4/06	3,455.00-dep	4/22	534.00
4/10	8.31	4/25	109.13-dep

Using the information on the partial bank statement given above:

a. Add the checks processed by the bank and record the total. __________

b. Add the deposits processed by the bank and record the total. __________

3. An office building is offering office space for rent. The manager of the building has the following spaces for rent: 323 square feet on the 1st floor, 257 square feet on the 2nd floor, 437 square feet on the 1st floor, 377 square feet on the 1st floor, 660 square feet on the 1st floor, and 470 square feet on the 2nd floor. Estimate, to the nearest hundred, the total number of square feet available on the 1st floor. __________

4. Computer Supply Wholesale Co. divides the state into four sales districts. Below is a partial list of customers by account number, district, year-to-date sales, and the amount owed on each account.

Customer Number	Sales District	Year-to-Date Sales	Current Balance
0012	C	$ 4,036.51	$ 1,455.21
0013	A	1,244.56	-0-
0024	C	35,036.42	30,840.03
0045	B	1,850.13	900.13
0132	D	215.00	100.11
0145	B	6,542.96	6,542.96
0153	C	2,711.10	71.10
0166	A	13,620.46	603.15
0168	C	8,797.75	4,220.49
0179	D	378.45	378.45

a. Find the total year-to-date sales. __________

b. Find the total current balance. __________

c. Find the total year-to-date sales for each district.

District A __________

District B __________

District C __________

District D __________

d. Verify your totals: The individual division totals should equal the total year-to-date sales. __________

unit 2

Subtraction of Whole Numbers and Decimals

"A good angle from which to approach any problem is the TRY-angle."

Addition and subtraction operations are used in calculations on all business financial papers—bank records, tax reports, sales invoices, payroll records, and business-related reports. You should be able to apply addition and subtraction processes with understanding, accuracy, and speed, both mentally and on the calculator. These math skills will help you in most jobs, and they may help you obtain promotions in business.

After studying this unit, you should be able to:

1. subtract whole numbers and decimals and verify differences
2. subtract to obtain a negative or credit balance
3. use the calculator to subtract whole numbers and decimals
4. obtain subtotals during addition and subtraction of whole numbers and decimals
5. use subtraction to solve business math problems

MATH TERMS AND CONCEPTS

The following terminology is used in this unit. Become familiar with the meaning of each term so that you understand its usage as you develop math skills.

1. **Borrowing**—increasing the value of a digit by taking one unit from the place value to its left.
2. **Difference**—the number obtained by subtracting one number from another—the answer.
3. **Minuend**—the top or first number in a subtraction problem.
4. **Negative/Credit Balance**—the difference when a larger number is subtracted from a smaller number.
5. **Subtraction**—the process of finding the difference between two numbers; the reverse of addition.
6. **Subtrahend**—the number being subtracted from the minuend.

Subtraction Process

Subtraction is the process of finding the difference between two numbers (**minuend** and **subtrahend**). It is the inverse of addition. In fact, subtraction *undoes* what addition does.

Example:

$$\begin{array}{r l} 15 & \text{minuend} \\ -10 & \text{subtrahend} \\ \hline 5 & \text{difference} \end{array} \qquad \begin{array}{r l} 5 & \text{addend} \\ +10 & \text{addend} \\ \hline 15 & \text{sum} \end{array}$$

To *verify* the accuracy of subtraction, add the subtrahend and the **difference.**

Example:

Subtract

$$\begin{array}{r l} 15 & \text{minuend} \\ -10 & \text{subtrahend} \\ \hline 5 & \text{difference} \end{array}$$

Verify

$$\begin{array}{r} 15 \\ -10 \\ \hline 5 \\ +10 \\ \hline 15 \end{array}$$

Subtracting zero (0) from a number does not change the value of that number.

Examples:

$$\begin{array}{r} 7 \\ -0 \\ \hline 7 \end{array} \qquad \begin{array}{r} 8 \\ -0 \\ \hline 8 \end{array}$$

Is your calculator turned off?

In the following exercises, fill in the blank with a number that will complete the number sentence.

1. 8 + 3 − 1 = ________

2. 19 = ________ − 3

3. 35 + 34 − 24 = ________

4. 30 = ________ − 15

5. 9 − 9 + ________ = 8

6. 89 − ________ = 39

7. 271 − 0 = ________

8. 16 − ________ = 16

9. 30 − ________ = 30

10. ________ − 0 = 33

Now find the difference in each of these problems.

11. $\begin{array}{r} 17 \\ -\ 7 \\ \hline \end{array}$ **12.** $\begin{array}{r} 92 \\ -12 \\ \hline \end{array}$ **13.** $\begin{array}{r} 399 \\ -\ 89 \\ \hline \end{array}$ **14.** $\begin{array}{r} 695 \\ -\ 84 \\ \hline \end{array}$ **15.** $\begin{array}{r} 815 \\ -214 \\ \hline \end{array}$

16. $\begin{array}{r} 3{,}009 \\ -1{,}007 \\ \hline \end{array}$ **17.** $\begin{array}{r} 1{,}745 \\ -1{,}434 \\ \hline \end{array}$ **18.** $\begin{array}{r} 579 \\ -262 \\ \hline \end{array}$ **19.** $\begin{array}{r} 2{,}797 \\ -\ 360 \\ \hline \end{array}$ **20.** $\begin{array}{r} 251 \\ -\ 40 \\ \hline \end{array}$

Borrowing

When subtracting numbers containing two or more digits, it may be necessary to borrow from a number in the next column.

Example:

$$\begin{array}{r} 72 \\ -28 \\ \hline \end{array} \qquad \begin{array}{r} \overset{6}{\cancel{7}}\,{}^{1}2 \\ -2\ 8 \\ \hline 4\ 4 \end{array}$$

The 7 becomes 6 and the 2 becomes 12.

Three-digit numbers may require that you borrow twice.

Learn to borrow mentally!

Example:

$$\begin{array}{r} 572 \\ -188 \\ \hline \end{array} \qquad \begin{array}{r} \overset{4}{\cancel{5}}\,{}^{1}\overset{6}{\cancel{7}}\,{}^{1}2 \\ -1\ 8\ 8 \\ \hline 3\ 8\ 4 \end{array}$$

The 7 becomes 6 and the 2 becomes 12; the 6 becomes 16 and the 5 becomes 4.

Do not use your calculator!

Find the difference in each of these problems. Work the problems as quickly as you can.

* *Remember, if you must carry or borrow, do so mentally.*

1.	223 − 78	**2.**	6,640 −3,960	**3.**	134,978 − 24,879	**4.**	9,473,200 −2,830,923	**5.**	1,000 − 867
6.	354 − 97	**7.**	2,680 −2,598	**8.**	1,023 − 685	**9.**	7,921 −4,853	**10.**	6,002 −5,773

Vertical and Horizontal Subtraction

The decimal point is aligned vertically for ease in subtracting. Align and subtract in each of the following problems. Add zeros where needed for ease in alignment.

1. 584.862 − 22.5

2. 789 − 333.5

3. 37.49 − 29

4. 310.89 − 17.543

5. 49.47 − 47.1 = ________

6. 267.983 − 185.76 = ________

Negative or Credit Balances

When a larger number is subtracted from a smaller number, the difference is a **negative balance**.

Example:

64.90	minuend	45.26	minuend
−45.26	subtrahend	−64.90	subtrahend
19.64	difference	−19.64	negative balance

Negative balances can be indicated in one of these ways:

1. *Negative sign* before or after a number: −19.64 or 19.64−.
2. *Parentheses:* (19.64). Used frequently in checkbooks, accounting, and indicating decreases in comparing figures in one time period to an earlier time period.
3. *CR* after a number: 19.64 CR. CR stands for **credit balance** and is sometimes used in business.

To verify the accuracy of subtraction, ignore the negative sign and add the minuend and the difference.

Example:

Subtract	Verify
45.26	45.26
(64.90)	(64.90)
(19.64)	19.64
	+45.26
	64.90

* *Recording negative balances in red ink is no longer practical because so many records are reproduced on equipment that does not duplicate colors.*

Find the credit balance in each of the following problems. Use *parentheses* to indicate a credit balance.

1.	300.00 −600.00 (300.00)	**2.**	263.40 −446.20	**3.**	188.35 −243.89	**4.**	29.94 −39.84	**5.**	64.50 −139.27
6.	247.18 −473.00	**7.**	24.63 −29.72	**8.**	34.66 −705.21	**9.**	936.00 −998.00	**10.**	785.25 −1,874.50

Practice

Find the difference in each of the following problems.

* ***Remember to indicate credit balances if required.***

1.	80,172,624 −53,997,758	**2.**	14,025.27 − 7,352.04	**3.**	432,039.74 −277,050.72
4.	218,774.33 − 40,695.10	**5.**	7,003.79 −1,008.79	**6.**	65,042.333 − 631.898
7.	6,064 −2,949	**8.**	13,271 − 3,678	**9.**	9,000 −4,325
10.	28,412 −17,562	**11.**	3,785 −7,123	**12.**	9,736 −11,012

13. 5,846 − 2,527 = ________ **14.** 213,873 − 256,004 = ________

Did you verify your answers?

CALCULATOR INSTRUCTIONS

* ***Remember to refer to Appendix A, page 335, for specific calculator instructions.***

Locate the following:

Minus Key

Subtotal Key

Solve the problem: 456 − 348

1. Set the decimal selector on 0.
2. Check your posture.
3. Clear your calculator.
4. Enter 456 by touch and operate the plus key.
   * ***For operation of pocket calculators, refer to the manual.***
5. Enter 348 by touch and operate the minus key.
6. Obtain a total by operating the total key.
   Did you get 108 as the difference?

Practice

Eyes on your book!

Subtract each of the following problems. * ***Set decimal selector on 0.***

1.	2.	3.	4.	5.
387	751	375	963	660
−348	− 48	−109	−502	− 34
39				

6.	7.	8.	9.	10.
1,431	4,870	8,917	5,827	1,936
− 259	− 971	−4,106	−3,243	− 159

Credit Balances

Credit balances appear in red, are followed by a minus sign, or appear in parentheses on printing calculators. Displays indicate a credit balance by a light, a negative symbol, or parentheses.

* ***Using parentheses to indicate negative numbers is the most common method in business.***

Decide which form to use to indicate credit balances.

Indicate credit balances as you record the answers to the following problems.

1.	2.	3.	4.	5.
197	106	7,106	753	6,632
−355	−587	−8,336	−861	−9,083

6.	7.	8.	9.	10.
3,240	2,091	30,844	8,624	45,800
−5,800	−2,784	−31,975	−11,389	−52,638

Subtraction of Decimals

To subtract numbers with two decimal places, set the calculator's add mode and decimal selector if required. If you do not have the add mode position, set the decimal selector on 2. If the problem has varying decimals, set the decimal selector to accommodate the number with the most decimals and enter the decimal point manually.

Find the difference in each of the following problems.

1.	2.	3.	4.
16.28	687.04	37,421.60	366.73
− 4.51	−895.22	− 9,671.53	−334.98

5.	6.	7.	8.
759.26	3,956.442	3,090.14	773.32
−817.48	−1,464.70	− 649.601	−873.64

Did you indicate credit balances correctly?

Addition and Subtraction of Decimals with Subtotals

The *subtotal key* is used to print an accumulated total without clearing the total from the calculator. Displays show the subtotal on the screen after each entry. On printing or display/printing calculators, the subtotal is indicated by a symbol on the printout tape when the subtotal key is used.

In problems 1-10, record both the subtotal and the total.

* *Set your calculator's add mode or set the decimal selector on 2.*

1.
22.79
35.00
−79
−21.21 S
52.50
−94.68
(63.39) T

2.
121.00
415
−384.66
S
368.00
−67.80
T

3.
88.79
−218.60
−39
S
645.30
16.00
T

4.
.45
1.27
.76
.35
S
.51
2.90
.74
.86
T

5.
88.70
.46
−78.98
248.93
S
2.67
−78.72
74.65
81.93
T

6.
.84
6.47
−1.20
.55
S
−4.97
−.49
7.16
−2.24
T

7.
25.43
−6.75
29.81
−149.71
92.35
S
35.78
−.21
−24.40
50.00
T

8.
734.57
−767.40
−97.34
1,599.76
S
−299.42
34.87
8.63
−5.56
T

9.
47.70
−49.20
92.30
69.40
−39.42
−10.87
73.24
35.85
12.19
40.69
S
9.70
−93.15
66.35
−2.10
−74.52
7.52
30
50.18
6.70
18.30
T

10.
25.65
−5.21
69.36
−56.50
−320.10
13.99
−20.95
2.90
7.65
40.32
S
60.30
−111.00
24.82
−17.00
99.50
−8.04
11.52
T

Did you indicate credit balances correctly?

EVALUATING YOUR SKILLS

You will be timed on the following problems. Your minimum goal is to complete the problems in 13 minutes. Work each problem once.

1. 3.76, −1.89
2. 7.22, −5.41
3. 8.50, −7.48
4. 6.91, −3.18
5. 5.18, −4.19

6. 2.17, −3.56
7. 932.12, −175.34
8. 146.38, −182.37
9. 942.93, −744.64
10. 651.47, −818.43

11. 34.45, −55.33
12. 94.68, −12.67
13. 60.32, −51.68
14. 72.40, −28.56
15. 5.84, −3.96

16. 16.86, 64.13, −35.52, 89.80
17. 45.00, −25.38, 81.61, −41.86
18. 383.00, 768.21, −244.24, 629.13
19. 238.42, −657.10, −942.64, 360.44
20. 653.06, −159.97, 577.11, −343.50

21.	22.	23.	24.	25.
175.81	564.07	5,912.34	4,571.31	3,751.74
824.17	−552.99	−3,032.80	−4,057.00	8,403.55
−274.00	394.52	6,226.08	−7,459.76	7,311.48
−255.30	−840.47	2,246.16	1,220.88	−6,440.79
943.85	458.40	7,026.85	1,593.52	−3,581.28
678.60	857.51	−1,335.15	−5,613.03	5,524.56
596.25	448.31	5,116.77	3,812.11	3,141.54
		9,263.40	−7,882.18	3,588.53
			3,526.88	1,646.41
			3,000.45	2,240.64

If you were unable to meet your goal, or if you want to set your goal higher, practice the problems on pages 21 and 22 to develop speed and accuracy. Then repeat the problems above.

Grading Scale

Completion Time	0 Errors	1 Error	2 Errors	3 Errors
9 minutes	A+	A	B+	B
9½ minutes	A	B+	B	C+
10 minutes	B+	B	C+	C
10½ minutes	B	C+	C	C−
11 minutes	C+	C	D+	D
11½ minutes	C	D+	D	D−
12 minutes	D+	D	D−	repeat
12½ minutes	D	D−	repeat	
13 minutes	D−	repeat		

Grading Scale—End of 5 Weeks

Completion Time	0 Errors	1 Error	2 Errors	3 Errors
7½ minutes	A+	A	B+	B
8 minutes	A	B+	B	C+
8½ minutes	B+	B	C+	C
9 minutes	B	C+	C	D+
9½ minutes	C+	C	D+	D
10 minutes	C	D+	D	repeat
10½ minutes	D+	D	repeat	

Grading Scale—End of 9 Weeks

Completion Time	0 Errors	1 Error	2 Errors	3 Errors
6 minutes	A+	A	B+	B
6½ minutes	A	B+	B	C+
7 minutes	B+	B	C+	C
7½ minutes	B	C+	C	D+
8 minutes	C+	C	D+	D
8½ minutes	C	D+	D	repeat
9 minutes	D+	D	repeat	

APPLYING YOUR KNOWLEDGE

Using the information given, calculate the correct answers. Some problems may require you to use the metric conversion chart in Appendix B, page 339. * *Set the decimal selector on add mode or 2 unless otherwise directed.*

1. Dena's job as a salesperson for a gift store requires her to make change. A customer buys four items costing $2, $5.25, $3.47, and $6.50. The sales tax is 87¢. How much change will the customer receive if the customer gives Dena a $20 bill?

____

2. A manufacturer of precision instruments set guildelines for a piece of metal to measure from a minimum of .9998 inches to a maximum of 1.003 inches in length. What is the difference between the minimum length and the maximum length? * *Set decimal selector on 4.*

____

3. The manufacturer in Problem 2 wants to export the metal part and must use metric measures in millimeters. What is the difference between the minimum length in millimeters and the maximum length in millimeters? * *Set decimal selector on 4.*

____

4. Jerri works for a local garbage collector. She received her paycheck on October 1 in the amount of $864.75. During the month of October she paid the following:

Car Payment	$395.50
Gas and Oil	75.19
Car Insurance Payment	67.00
Clothing	54.02
Compact Discs	32.89
Movies	15.00
Food	123.50
Savings Account	75.00

a. What were Jerri's total car expenses for October?

____

b. What were Jerri's total expenses for October?

____

c. How much was left from Jerri's October paycheck?

____

5. If David sold $2,789 in men's jewelry this week and $2,073 last week, how much more were this week's sales than last week's?

____

6. An invoice from XYZ store for office supples totaled $379.23. A check was written for that amount. An error was found on the invoice and the total should have been $337.89. How much money should the store refund?

____

7. Henry and Johnson, Inc. bought a building in 1964 for $39,000 and sold it in 1990 for $219,000. What was the gain from the sale?

____

8. Angela has started a business at home to earn extra money while in school. The business had earnings of $95 the first week. Her expenses for the week were as follows: Paper—$22.50, Disks—$3.00, Photocopies—$1.80, and Electricity—$10.00. On the budget form, record her actual costs and determine if she stayed within her budget.

a. CASH BUDGET

Item	Cash: Budgeted	Cash: Actual
Computer Disks	$ 9.00	______
Photocopy Work	1.00	______
Paper	27.95	______
Electricity	8.50	______
Total Cash Budgeted and Spent	$46.45	______
Amount of difference between Actual and Budgeted		______

b. How much did she earn after her expenses?

____

9. As of November 1, Petite Clothing Store owed $1,028 to Cloth Wholesale. On November 17, they bought five items costing $549.20 on credit. On November 19, they returned some items costing $58.34 and paid $800 on account. How much does Petite Clothing still owe Cloth Wholesale?

____

10. The United States consumed 17.7 kilograms of rice per individual from 1974 to 1987. During the same time, Japan had a per individual consumption of 78.8 kilograms. How much more did Japan consume than the United States?

____

ADVANCED APPLICATIONS

Using the information given, calculate the correct answers.

1. The Hutto Apparel Shop keeps a record of amounts owed, amounts charged, items returned, and payments for each charge customer. Determine how much each customer owes on February 28.

a. Felkins ------------------- 12-2 —Charge $ 35.89
1-2 —Charge $329.41
1-3 —Returned items $ 58.00
2-1 —Paid $215.00
2-28—Amount Owed __________

b. Henry --------------------- 12-2 —Charge $547.68
12-30—Paid $547.68
1-5 —Charge $ 24.73
1-22—Charge $ 46.00
2-28—Amount Owed __________

c. Schooley---------------- 12-30—Amount owed $238.53
12-30—Paid $200.00
1-17—Charge $ 78.93
1-18—Refund by store $ 50.00
2-28—Amount owed __________

d. Johnson----------------- 12-18—Amount owed $549.41
12-30—Charge $ 31.76
12-30—Returned items $153.80
1-2 —Paid $275.00
2-15—Charge $ 47.39
2-28—Amount owed __________

e. What is the total amount owed by the four customers on February 28? __________

f. What was the total amount of returned items? __________

2. A bank keeps a daily record of the balance in each account. Once a month, a statement is sent to each customer showing deposits and checks processed. Deposits are added to the balance and checks are subtracted. Find the cash balance for each date on this partial bank statement.

Checks	Checks	Checks	Deposits	Date	Balance
				3/29	$23,587.51
39.78				4/03	______
			581.92	4/04	______
350.00	2,113.00		1,742.93	4/07	______
79.64	153.48			4/08	______
20.56				4/09	______
131.38				4/10	______
7.99	37.15	64.64	593.77	4/13	______
58.59	71.36			4/14	______
242.38	462.71			4/14	______
61.58	69.14	149.66		4/15	______
157.31	192.47			4/16	______
			349.50	4/16	______
14.00	43.77	60.82	2,687.33	4/16	______
79.41	82.96	229.48		4/17	______
787.78				4/17	______
169.74	253.00			4/20	______
10.37				4/21	______

3. The number of telephones in the U.S. in 1986 was 76 for every 100 people. During the same time, Australia had 55 telephones for every 100 people, Mexico had 9.6, and Germany had 64. How many more telephones were there in the U.S. per 100 people than in Germany?

______

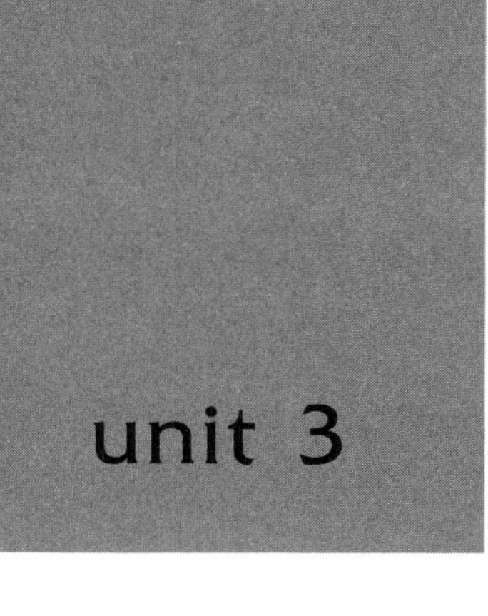

unit 3 Multiplication of Whole Numbers and Decimals

"Grumblers never work; workers never grumble."

Multiplication simplifies the process of many business calculations including sales tax, invoice extensions, payroll taxes, and percentages. Once you can perform multiplication both manually and mentally, you can use the calculator to shorten the process and assure accuracy. You will find some functions can be performed more quickly *without* the calculator. You will find shortcuts in the problem-solving process through accumulative multiplication, negative multiplication, and multifactor multiplication with the use of the calculator.

After studying this unit, you should be able to:

1. multiply whole numbers and decimals and verify products
2. multiply mentally by multiples of the number 10
3. round whole numbers and decimals
4. apply the procedure for rounding numbers to estimate answers in multiplication
5. multiply mentally by 1¢ and 10¢
6. multiply multifactor problems
7. use the calculator to multiply whole numbers and decimals
8. use the calculator to multiply with constants
9. combine multiplication and addition or subtraction using the accumulative operation on the calculator
10. solve business math problems using multiplication

MATH TERMS AND CONCEPTS

The following terminology is used in this unit. Become familiar with the meaning of each term so that you understand its usage as you develop math skills.

1. **Constant**—a number used more than once for the same operation.
2. **Constant Multiplication**—multiplication using the same multiplier for more than one problem.
3. **Extension**—an amount obtained by multiplying the quantity by the unit price. The symbol @ is often used to indicate multiplication.
4. **Factor**—any number being multiplied.
5. **Gross Profit**—for a merchandising business, sales (income) less the amount the business paid for the merchandise.

6. **Multiplicand**—a number that is multiplied by the multiplier (first number in a multiplication problem).
7. **Multiplication**—the operation of repeated addition.
8. **Multiplier**—the number of times by which the multiplicand is multiplied (second number in a multiplication problem).
9. **Price**—the amount a customer pays for merchandise.
10. **Product**—the number obtained by multiplying two or more numbers together.
11. **Profit**—the difference when the amount received is greater than the amount spent.
12. **Quantity**—the number of items purchased or sold.
13. **Sales (income)**—total amount received, less taxes, for items sold.
14. **Unit Price**—the price of a single item.

Multiplication Process

Multiplication is repeated addition. One number (**multiplicand**) is added to itself as many times as there are units in the other number (**multiplier**). In the problem 25 (multiplicand) × 17 (multiplier), the number 25 is added to itself 17 times. Multiplication shortens the process of addition. You first find 7 × 25, then 10 × 25, and add the answers to obtain the **product**.

Example:

$$\begin{array}{r} 25 \\ \times\ 7 \\ \hline 175 \end{array} \qquad \begin{array}{r} 25 \\ \times 10 \\ \hline 00 \\ +250 \\ \hline 250 \end{array} \qquad \begin{array}{r} 25 \\ \times 17 \\ \hline 175 \\ +250 \\ \hline 425 \end{array} = \text{Product}$$

The order in which numbers are multiplied does not affect the product.

Examples: $7 \times 6 = 42$ $\quad 3 \times 12 \times 4 = 144$
$6 \times 7 = 42$ $\quad 4 \times 3 \times 12 = 144$
$12 \times 4 \times 3 = 144$

* ***To verify multiplication, reverse the order in which the factors are multiplied. You should obtain the same product.***

When a number is multiplied by *one*, the value of that number will not change.

Examples: $7 \times 1 = 7$ $\quad 16 \times 1 = 16$

The product of any number multiplied by 0 will be 0.

Examples: $7 \times 0 = 0$ $\quad 16 \times 0 = 0$

Fill in the blank with a number that will complete the number sentence.

1. 52 × 78 = 78 × ________ **2.** 70 = 1 × ________

3. 213 × 1 = ________ **4.** 28 × ________ = 0

5. 49 × 0 = ________ **6.** If 2 × 3 × 7 = 42, then 7 × 3 × 2 = ________

Now find the product in each of the following problems.

Is your calculator turned off?

7. 68 × 9 = 612 **8.** 45 × 13 **9.** 73 × 45 **10.** 702 × 19 **11.** 783 × 384

12. 14 × 38 **13.** 875 × 645 **14.** 426 × 64 **15.** 381 × 528 **16.** 1,562 × 27

Multiplication by Multiples of 10

Multiplication of whole numbers by 10, 100, 1,000, or other multiples of 10 can be performed quickly without the calculator. When multiplying by multiples of 10, add the number of zeros in the multiplier to the right of the multiplicand. The result is the product.

Examples: 426 × 10 = 4,260 426 × 100 = 42,600

Find the products of the following.

Is your calculator turned off?

1. 32 × 10 = 320 **2.** 32 × 100 = ______

3. 167 × 100 = ______ **4.** 32 × 1,000 = ______

5. 167 × 1,000 = ______ **6.** 4,852 × 10 = ______

To multiply by 20, 30, 40, etc., simply multiply by the number to the left of the zero(s). Then add zeros to the right of the *product* for each zero in the multiplier.

Example: 426 × 2 = 852
426 × 20 = 8,520

Do the calculations *mentally* as you complete these problems.

7. 14 × 30 = ______ **8.** 75 × 500 = ______

9. 92 × 3,000 = ______ **10.** 15 × 70 = ______

11. 14 × 2,000 = ______ **12.** 88 × 300 = ______

Multiplication of Decimals

When multiplying numbers containing decimals, first find the product of the numbers being multiplied. Then find the *sum* of the number of decimal places in the multiplicand *and* multiplier. This sum is the number of decimal places in the product. Beginning at the right of the product count left and place the decimal point.

Example:

2.5	1 decimal place
×.13	+2 decimal places
75	
+25	
.325	3 decimal places

If there are more decimal places in the **factors** than there are digits in the product, add a zero(s) to the *left* of the product. Then place the decimal point to the left of the zero(s).

Example:

.25	2 decimal places	
×.13	+2 decimal places	
75		*Note a zero must be added to the left of the product.
+25		
.0325*	4 decimal places	

Place the decimal point in the correct position in the product in each of the following problems. Add zeros to the left of the product where necessary.

1.	2.	3.	4.	5.
12.5	.15	.09	439.5	597.6
×.143	× .6	×2.4	× 2.56	× 78.7
375	90	36	26370	41832
500		+18	21975	47808
×125		216	+8790	+41832
17875			1125120	4703112

When multiplying decimals by multiples of 10, simply move the decimal point one digit to the right in the multiplicand for every zero in the multiplier. Add a zero to the *right* of a number if an additional place value is needed.

Examples:

426.52 × 10 = 4,265.2
426.52 × 100 = 42,652. = 42,652
426.52 × 1,000 = 426,520. = 426,520*

*Note a zero was added.

Calculate the products for the following problems.

Is your calculator off?

6. 3.29 × 1,000 = ____________
7. 8.35 × 100 = ____________
8. 10 × 518 = ____________
9. 9,529 × 1,000 = ____________
10. 42.01 × 1,000 = ____________
11. 23.388 × 10 = ____________

Rounding Numbers

A business will often use approximate numbers rather than exact figures. Financial statements and tax returns can be prepared with all amounts rounded to the nearest hundreds or thousands place. Deciding on how many places to round a number depends on how accurate an answer is needed.

Once you decide to which place value the number is to be rounded, look at the digit to the right of that place value. *If the digit is 5 or greater,* add 1 to the figure in the place value you are rounding to and change all digits to the right of this place value to zero. *If the digit to the right is 4 or less,* change it and all digits to the right to zero.

Examples:

1,535 rounded to nearest hundred is 1,500
1,500 rounded to nearest thousand is 2,000

Round the following whole numbers as indicated.

1. 1,373 rounded to nearest hundred = ____________
2. 543,169 rounded to nearest thousand = ____________

3. 345 rounded to nearest ten = ______

4. 6,455 rounded to nearest thousand = ______

5. 337,162 rounded to nearest ten thousand = ______

You can also round numbers when all the decimal places in an answer are not needed. For example, if an answer has three decimal places when only two are needed, look at the third digit to the right of the decimal point. If the third digit is 5 or greater, add 1 to the second digit to the right of the decimal point. If the third digit is 4 or less, drop the third digit.

Examples: .453 is rounded to .45
.457 is rounded to .46

Round the following to the nearest cent.

6. $13.568 = ______ 7. $97.375 = ______

8. $12.896 = ______ 9. $78.972 = ______

10. $10.423 = ______ 11. $44.6445 = ______

Sometimes a situation may require that you round to the nearest whole number (dollar). If the first digit to the right of the decimal point is 5 or greater, add 1 to the first digit to the left of the decimal point. If the first digit to the right of the decimal point is 4 or less, drop the number(s) to the right of the decimal point.

Examples: $14.16 is rounded to $14
$14.57 is rounded to $15

Round the following to the nearest dollar.

12. $13.15 = ______ 13. $7.34 = ______

14. $244.61 = ______ 15. $1,364.49 = ______

16. $349.92 = ______ 17. $158.43 = ______

Estimating Answers

The procedure for rounding can be used to estimate answers. In the problem 235 × 41 = 9,635, the estimated answer is 9,600 (240 × 40). The estimated answer tells you if your actual answer could possibly be correct.

Give the *estimated* answer for each of the following problems. Round the multiplicand and the multiplier to the nearest ten.

1. 448 × 44 = 18,000 2. 236 × 78 = ______

450 × 40

3. 41 × 72 = ______ 4. 427 × 22 = ______

5. 167 × 12 = ______ 6. 49 × 97 = ______

Now multiply to see if your estimated answers are close to the actual answers.

7. 448×44 **8.** 236×78 **9.** 41×72 **10.** 427×22 **11.** 167×12 **12.** 49×97

* ***Use the estimating method to check your answers when multiplying both mentally and with the calculator.***

Multiplication by 1¢ and 10¢

To multiply a number by 1¢ (.01), move the decimal point two places to the left in the product. To multiply by 10¢ (.10 = .1), move the decimal point one place to the left in the product.

Examples: 45 × .01 = .45
45 × .10 = 4.50

Mentally solve these problems.

1. 23 × .01 = ____________ **2.** 487 × .10 = ____________

3. 37 × .01 = ____________ **4.** 310 × 1¢ = ____________

5. 337 × 10¢ = ____________ **6.** 591 × 10¢ = ____________

Practice

These problems give practice in multiplication by 10, 100, 1,000, .01, .1, and multiples of 10. Work the problems *mentally* as quickly as you can.

Is your calculator off?

1. 72.4 × 100 = ____________ **2.** 8.392 × .01 = ____________

3. 12 × 30 = ____________ **4.** 9.43 × 1,000 = ____________

5. 244.8 × .01 = ____________ **6.** 711 × .10 = ____________

7. 782 × 10 = ____________ **8.** .529 × 100 = ____________

9. 13 × 40 = ____________ **10.** 400 × 3.0 = ____________

Multifactor Multiplication

Multifactor multiplication is multiplying three or more numbers in one problem. Begin multiplying with the first factor on the left, and find a product for the first two factors. Then multiply that product by the third factor. Continue this multiplication process until you have multiplied all the numbers in the problem.

Example: 2 × 3 × 4 =
2 × 3 = 6
6 × 4 = 24

Find the products in the following problems.

1. 3 × 6 × 8 = 144 **2.** 16 × 4 × 3 = ____________

3 × 6 = 18
18 × 8 = 144

3. $5 \times 2 \times 7 =$ ____________ 4. $11 \times 10 \times 6 =$ ____________

5. $3 \times 4 \times 5 \times 7 =$ ____________ 6. $42 \times 7 \times 1 =$ ____________

7. $80 \times 1 \times 4 =$ ____________ 8. $5 \times 11 \times 3 =$ ____________

CALCULATOR INSTRUCTIONS

Locate the following:

Multiplication Key

Rounding Selector—Calculators equipped with a rounding selector will have two or three positions. The 0 position (or ↓) simply cuts an answer off at the number of decimal places set. The 5/4 or 5 position will round numbers following the rounding rules on pages 32 and 33. The 9 (or ↑) position rounds everything up.

Floating Position on the Decimal Selector—Represented by F or FL. This allows the keying of numbers with varying decimal places resulting in an answer being carried out without rounding or cutting off the decimal places.

Constant Position—Most calculators have a constant position built in. A few require a constant position to be set (K).

Accumulator—Use the grand total (GT), accumulator (ACC), or sigma position (Σ).

Solve the problem: $247 \times 16 =$

* *Enter the problem into the calculator exactly as it is read.*

1. Set the decimal selector on 0.
2. Check your posture.
3. Clear the calculator.
4. Enter 247 and operate the multiplication key.
5. Enter 16 and operate the equals key.
   Did you get 3,952?

Practice

Using your calculator, find the product of each problem below.

* *Set decimal selector on 0.*

1. 24 × 46 = 1,104
2. 27 × 91
3. 264 × 67
4. 855 × 29
5. 302 × 13
6. 799 × 123
7. 218 × 200
8. 45,971 × 823
9. 9,340 × 255
10. 10,666 × 38

Multiplication of Decimals

Enter the decimal point manually for all multiplication problems. Remember, the add mode is used only for addition and subtraction of two decimals.

When you need more decimal places than your calculator settings allow, including addition and subtraction, set the selector on F or FL.

When the maximum number of decimals is needed in a product, the decimal selector is set on floating. Set the decimal selector on F or FL to complete the following problems. Do not round answers.

* ***Remember to enter the decimal point manually.***

1.	3.21 × .14 .4494	**2.**	26.8 × 80	**3.**	64 × .7	**4.**	3.664 × .25	**5.**	.16 × 2.23
6.	17.32 × 2.5	**7.**	18.1 × 40	**8.**	388.57 × .76	**9.**	368 × 3.8	**10.**	556.4 × 73.8

Now work the next 10 problems rounding each product to two decimal places. Move the decimal selector to 2. Enter all decimal points into the calculator.

* ***Calculators with rounding selectors must be set on the 5/4 position to round. Refer to your calculator manual for specific instructions on rounding features.***

Is the rounding selector on 5/4 and the decimal selector on 2?

11.	66.35 × 3.2	**12.**	406.14 × 3.5	**13.**	200 × .2	**14.**	228.75 × 17.43	**15.**	7.152 × 2.95
16.	1.41 × 7.66	**17.**	23.56 × 6.28	**18.**	321.58 × 58.17	**19.**	234.102 × 11.22	**20.**	742 × .82

Multifactor Multiplication

Solve the problem: 37 × 4 × 20 =

* ***Enter the problem into the calculator exactly as it is read.***

1. Set the decimal selector on 0.
2. Check your posture.
3. Clear the calculator.
4. Enter 37 and operate the multiplication key.
5. Enter 4 and operate the multiplication key.
6. Enter 20 and operate the equals key.
   Did you get 2,960?

* ***Some calculators require a product to be taken after each series. The product must then be reentered into the calculator by operating the multiplication key.***

Remember the decimal selector is set depending on the problem. Multiplying whole numbers requires 0 decimal places. Rounding to two places requires 2 decimal places.

Find the product for each of the following problems.

* ***Round answers to two decimal places.***

1. 31 × 42 × 7 = ____________ **2.** 6.4 × 53 × 8 = ____________

3. 3.67 × 4 × 21 = ____________ **4.** 11 × .21 × 675 = ____________

5. 82 × .7 × 2 = ____________ **6.** 53 × .9 × 2.5 = ____________

Constant Multiplication

Constant multiplication allows you to enter a number (**constant**) once to solve a series of multiplication problems that include the same number (constant).

Example: $5 \times 4 = 20$
$5 \times 2 = 10$
$5 \times 3 = 15$

On some calculators you need to enter the constant first while on other models you will enter the constant second. To determine if your calculator uses the first or second entry as a constant, refer to Appendix A, page 335.

Solve the problem: 6×12
6×8
6×22

For calculators with the *first* entry as the constant:

1. Set the constant mode if necessary.
2. Set the decimal selector on 0.
3. Check your posture.
4. Clear the calculator.
5. Enter 6 and operate the multiplication key.
6. Enter 12 and operate the equals key.
   Did you get 72?
7. Enter 8 and operate the equals key.
   Did you get 48?
8. Enter 22 and operate the equals key.
   Did you get 132?

For calculators with the *second* entry as the constant:

1. Repeat Steps 1-4 above.
2. Enter 12 and operate the multiplication key.
3. Enter 6 and operate the equals key.
   Did you get 72?
4. Repeat Steps 7-8 above.
   Did you get 48 and 132?

Work problems 1–4 using the constant multiplication features of your calculator. * ***Round answers to two decimal places.***

Remember to set your decimal selector on 2.

1. $1.14 × 40 = ________

$1.14 × 38 = ________

$1.14 × 36 = ________

$1.14 × 9 = ________

2. $3.14 per dozen × 216 = ________

$3.14 per dozen × 187 = ________

$3.14 per dozen × 325 = ________

$3.14 per dozen × 103 = ________

3. 3.5 × 5.04 = ______

24.7 × 5.04 = ______

86 × 5.04 = ______

19.22 × 5.04 = ______

4.

Hourly Rate	×	Hours Worked	=	Total Pay
$5.15	×	39	=	________
$5.15	×	25	=	________
$5.15	×	14	=	________
$5.15	×	6	=	________

Accumulative Multiplication

Accumulative multiplication combines multiplication and addition or subtraction. This process can be done on most calculators. The result is a sum or grand total of products from a series of multiplication problems accumulated in the grand total register (GT), accumulative register (ACC), or memory register (M*). Many memory registers will have a sigma dial (Σ) or selector that will accumulate products into the memory bank.

* ***Refer to Appendix A, page 335, for how to use the sigma position.***

Solve the problem:

$$\begin{array}{r} 5 \times \$10.50 \\ 12 \times \$\ 6.95 \\ \underline{+\ 9 \times \$11.95} \end{array}$$

1. Set the decimal selector on 2.
2. Set the calculator so that it will accumulate.
3. Clear all registers.
4. Multiply 5 × $10.50 and do *one* of the following:
   a. operate the equals/plus key or equals key (calculators with accumulator).
   b. operate the equals key and/or memory plus key (memory bank models).

   Did you get $52.50?
5. Repeat the procedure in multiplying 12 × $6.95 and 9 × $11.95. Did you get $83.40 and $107.55?
6. Take a grand total by operating either the accumulator total key, the total key twice, or the memory total key.
   Did you get $243.45?

* ***Be sure all registers are cleared before starting a new problem.***

Use accumulative and constant multiplication to solve the following problems. * ***Round solutions to two decimal places.***

1. 1.54 × 37 = ____________
2.63 × .81 = ____________
3.61 × 7 = ____________
Total ____________

2. 159 × 1.78 = ____________
35.5 × 60 = ____________
742 × .72 = ____________
Total ____________

3. 35.4 × 2.87 = ____________
.612 × 46 = ____________
123 × 8.87 = ____________
Total ____________

4. 6.7 × 33 = ____________
.75 × 15 = ____________
2.02 × .38 = ____________
Total ____________

5.

Week	Hourly Rate		Hours Worked		Total Pay
1	$5.25	×	40	=	____________
2	$5.25	×	37	=	____________
3	$5.25	×	39	=	____________
4	$5.25	×	29	=	____________
	Total Pay Earned in Four Weeks				____________

Did you use constant and accumulative multiplication?

EVALUATING YOUR SKILLS

You will be timed on the following problems. Your goal is to complete all problems within 10 minutes. Your grade will be determined by the number of correct answers. Use your calculator and round answers to two decimal places. * *Set your decimal selector on 2.*

Whole Numbers

1. 45 × 34 = ____________

2. 138 × 201 = ____________

3. 74 × 98 = ____________

4. 368 × 157 = ____________

Decimals

5. 74.2 × 13.5 = ____________

6. 60.23 × 59.06 = ____________

7. 29.1 × 74.82 = ____________

8. 61.8 × 31.6 = ____________

Multifactor Multiplication

9. 13 × 53 × 58 = ____________

10. 40 × 60 × 658 = ____________

11. 79 × 200 × 493 = ____________

12. 16.1 × 78 × 74.1 = ____________

Constant Multiplication

13. 31 × 3.45 = ____________

14. 46 × 3.45 = ____________

15. 51 × 3.45 = ____________

16. 40 × 3.45 = ____________

17. 49 × 24.70 = ____________

18. 49 × 45.86 = ____________

19. 49 × 13.92 = ____________

20. 49 × 66.87 = ____________

Accumulative Multiplication

21. 46 × 14 = ____________

22. 68 × 43 = ____________

23. 36 × 26 = ____________

24. 51 × 30 = ____________

25. Total ____________

26. 67 × 381 = ____________

27. 43 × 75 = ____________

28. 40 × 229 = ____________

29. 72 × 61 = ____________

30. Total ____________

If you did not get at least 24 correct answers, go back and practice the problems in this unit. Then repeat the problems above.

Grading Scale

Grade	Number of Correct Answers
A	27-30
B	24-36
C	21-23
D	18-20

APPLYING YOUR KNOWLEDGE

Use the information given to calculate the answers. * *Remember to round answers to two decimal places. Read the problems carefully!*

1. The Lyrical Warehouse sold the following items. Extend each item listed to obtain the total sales for each.

	Total Sales
15 video tapes @ $24.95 each	________
57 cassette tapes @ $11.95 each	________
24 compact discs @ $15.99 each	________
14 posters @ $3.47 each	________
23 blank cassettes @ $3.25 each	________

2. The manager of a local restaurant purchased 7 sides of beef weighing an average of 256 lb each at $1.85 per lb. What was the total purchase price? ________

3. The Media Ticket Agency is selling tickets to a concert. The ticket teller was given 550 twenty-five dollar tickets and $75 cash to make change. At the end of the day the teller had 127 tickets and $10,624 in cash.
   a. How many tickets did he sell? ________
   b. How much money should he have collected? ________
   c. How much money should he have turned in? ________
   d. Did he turn in the correct amount of money? ________
   e. If not, how much more or less should he have turned in? ________

4. The Timely Grocery Store purchased 17 boxes of apples at $4.65 a box. Each box contained 4 dozen apples. They were sold for 15¢ each.
   a. What were total sales if all apples were sold? ________
   b. What was the gross profit?

      (Gross Profit = Sales − Amount Paid for Merchandise) ________

c. If a total of 18 apples had to be thrown away due to spoilage, what would be the total sales?

____

d. What would be the gross profit?

____

5. If a store sells 98 liters of soft drink in one day and the wholesale cost is 27¢ per gallon, what is the total cost of the soft drinks sold? (See metric conversion chart.)

____

6. An office building contains 78 offices. If 36 offices each rent for $1,345 per month and 42 offices each rent for $12,600 per year, what is the total rental income for a year?

____

7. Jeff Conine owns a modeling agency. The gross profit for this year was $943,561. The forecast for the gross profit for next year is 1.45 times this year's profit. What is the expected amount of gross profit for next year?

____

8. The modeling agency in Problem 7 employs 4 people full time and calculates their wages on an hourly basis. The remainder of the people needed are temporary employees and are paid by the job. Using the following information, determine the weekly earnings for each full-time employee.

a.

Employees	Hours Worked	×	Hourly Rate	=	Weekly Earnings
A. Jackson	40	×	$ 75.50	=	__________
B. Kelley	35	×	$ 67.75	=	__________
S. Henry	35	×	$ 80.00	=	__________
M. Lyons	40	×	$105.25	=	__________

(Weekly Earnings = Hours Worked × Hourly Rate)

b. If the same amount of wages are paid each week next year, how much of the expected gross profit will be left to cover other expenses after deducting the wages?

____

ADVANCED APPLICATIONS

Using the information given, calculate the correct answers.

1. Cincinnati Technical Center has determined that the cost of producing a letter is too high for its office. The president has been given a proposal to purchase a $12,000 computer to be used only for word processing. You have been asked to prepare the statistics to show if this new equipment would reduce letter production cost. Calculate the cost under the present system and the cost with the new equipment using the information below. Use 22 working days in a month and 12 months in a year. * *Set decimal selector on four.*

Cost per Letter Under the Present System	Estimated Cost per Letter Using the Computer
Dictator's Time— 8 min @ $.29565 per minute	Dictator's Time— 8 min @ $.29 per minute
Secretary's Time— 19 min @ $.1675 per minute	Secretary's Time— 10 min @ $.1675 per minute
Nonproductive Labor—$.648	Nonproductive Labor—$.42
Materials—$.3105	Materials—$.3105
Mailing—$.6012	Mailing—$.288 (use FAX machine)
Fixed Charges—$2.6696	Fixed Charges—$2.67

a. If 115 letters are produced in one day, determine:

	Present System	Proposed System
Cost per letter	______	______
Cost per day	______	______
Cost per month	______	______
Cost per year	______	______

b. What is the cost difference between the two systems? (The cost of purchasing the equipment must be included in your calculations.)

____

c. Based on your calculations, should the president approve the proposal?

____

2. The State Capitol uses water bottles with paper cups as drinking water facilities for the legislators. The Speaker of the House has received numerous complaints about the mess of paper cups and cost of maintaining water bottle service. During this session the legislators are going to set aside funds to purchase six water fountains to replace the current system. As the Speaker's aide, you were asked to conduct a cost analysis of the current system and the savings possible using new units. Based on the study, you found the following:

> One hundred fifty-three representatives get drinks an average of 7 times per day. They use an average of 3.5 paper cups per visit at a cost of $.009 per cup. There are 16 working days per month for the 8-month sessions. Intangible costs for storage of paper cups, water bottle service, and electricity are one half of the three-year cost of using paper cups. The new drinking fountains would cost $345 per unit, plus $125 installation for each of the 6 units. Electricity will cost $5.40 per month for each of the 6 units.

Determine the following: * *Round to two decimal places at each step.*

a. Current System Costs

Total visits per day (all representatives)

____

Daily paper cup usage

____

Paper cup cost per day

____

Cost per year for cups

____

Three-year cost for cups

____

Intangible costs

____

Total three-year cost of using water bottles and paper cups

____

b. Proposed System Costs

Cost for six water fountains

____

Installation cost

____

Total cost for units and installation

____

Total three-year electrical cost

____

Total three-year cost for units

____

c. Current System Costs

____

Proposed System Costs

____

Savings over Three Years

____

3. The Sweet Shop received a shipment of Dutch chocolate from Holland. There were 12 decorative tins containing .75 kilograms each and 8 boxes containing 2 kilograms each of chocolate. The tin containers were sold for $4.50 each; the boxes sold for $10.50 each. The store paid $3.00 per kilogram plus a $7.50 import fee. What was the gross profit if all the chocolate was sold?

____

unit 4 Division of Whole Numbers and Decimals

"Success is the difference between the desire to be and the determination to be."

Business people frequently use division in setting prices, comparing one fiscal period with another, finding unit costs, and computing averages. To divide accurately, you must know the multiplication tables. A simple error in any mathematical operation, even in decimal placement, can cost a company money and time.

After studying this unit, you should be able to:

1. divide whole numbers and decimals and verify quotients
2. divide mentally by multiples of the number 10
3. apply the procedure for rounding numbers to estimate answers in division
4. divide mentally by 1¢ and 10¢
5. solve unit price problems using quantity grouping in extending prices
6. compute simple averages
7. use the calculator to divide whole numbers and decimals
8. use the calculator to divide in problems containing constants
9. use the calculator to compute simple averages
10. solve business math problems using division

MATH TERMS AND CONCEPTS

The following terminology is used in this unit. Become familiar with the meaning of each term so that you understand its usage as you develop math skills.

1. **Cost of Goods Sold**—the amount a business pays for the items sold during a given period of time.
2. **Dividend**—the number or quantity to be divided.
3. **Division**—the process of separating a number into parts; the reverse of multiplication.
4. **Divisor**—the number by which the dividend is divided.
5. **Dozen (doz)**—a quantity equal to 12 items.
6. **Gross (gr)**—12 dozen or 144 items.
7. **Hundredweight (CWT)**—a unit of weight equal to 100 pounds.
8. **Inventory**—a list of the items a business has available to sell.
9. **Net Profit**—income less expenses.

10. **Operating Expenses**—what it costs to operate a business—rent, utilities, salaries, postage, etc.
11. **Per C**—by the hundred.
12. **Per M**—by the thousand.
13. **Quotient**—the number (the answer) obtained when one number is divided by another.
14. **Ream (rm)**—a quantity of paper equal to 500 sheets.
15. **Remainder**—what is left undivided when one number is divided by another.
16. **Selling Price (sales price)**—the amount a customer pays for merchandise.
17. **Simple Average**—a single number used to represent a group of numbers.
18. **Ton**—a unit of weight equal to 2,000 pounds.
19. **Unit Price**—the price of a single item.

Division Process

Division is the process of finding how many times a number (the **divisor**) is a part of another number (the **dividend**). The divisor is repeatedly subtracted from the dividend until the difference is less than the *divisor*. Division undoes what multiplication does. The answer is called the **quotient**.

Division can be indicated in several ways:

```
   15
5)75
```

$\frac{75}{5} = 15$ $\qquad$ $75 \div 5 = 15$

In each way, the divisor is 5, the dividend is 75, and the quotient is 15.

In the problem $5\overline{)7{,}530}$, you are finding how many times 5 is contained in 7,530. Follow these steps in finding the quotient.

1. Find how many times the divisor will go into each number in the dividend.

   Example: $5\overline{)7{,}530}$

2. Write the number of times directly above the dividend as part of the quotient and multiply the number times the divisor. Then bring down the product.

   Example:

```
  ×1
5)7,530
  5
```

3. Subtract the product from the number above it in the dividend.

   Example:

```
  ×1
5)7,530
 −5
  2
```

4. Bring down the next number in the dividend and repeat Steps 1-3.

   Example:

```
  ×15
5)7,530
 −5
  25
 −25
```

5. Repeat Steps 1-4 until the problem is completed.

Example:

```
   1,506
5)7,530
 -5
  25
 -25
    3
   -0
    30
   -30
```

* ***Notice that the computations are brought down in each step and subtracted from a number in the dividend. Notice that both the quotient and digits brought down are aligned with the dividend.***

When a number is divided by *one*, the value of the number does not change.

Example: $15 \div 1 = 15$

Dividing a number by *itself* gives a quotient of 1.

Example: $15 \div 15 = 1$

Dividing *zero* (0) by any number gives a quotient of zero (0).

Example: $0 \div 5 = 0$

Dividing any number by *zero* (0) is not possible.

In the following exercises, fill in the blank with a number that will complete the number sentence.

Is your calculator turned off?

1. 23 ÷ ________ = 1 **2.** ________ ÷ 15 = 4

3. ________ ÷ 49 = 0 **4.** 66 ÷ ________ = 66

5. 0 ÷ 52 = ________ **6.** 6 = 24 ÷ ________

Now divide in each of the following problems.

Is your calculator turned off?

7. 2)250 **8.** 33)1,485 **9.** 5)330

```
   125
2)250
 -2
   5
  -4
   10
  -10
```

10. 6)456 **11.** 3)1,239 **12.** 8)10,040

* *To verify division, multiply the quotient by the divisor. The answer is the dividend.*

Example: 7,530 ÷ 5 = 1,506
1,506 × 5 = 7,530

Verify the division problems you just completed.

13. (#7) ______ × 2 = ______ **14. (#8)** ______ × 33 = ______

15. (#9) ______ × 5 = ______ **16. (#10)** ______ × 6 = ______

17. (#11) ______ × 3 = ______ **18. (#12)** ______ × 8 = ______

Are your answers the same as the dividends?

Remainders

When one whole number will not divide into another whole number *evenly*, the quotient will have a **remainder**.

Example:

$$\begin{array}{r} 5 \\ 3\overline{)17} \\ -15 \\ \hline 2 \end{array} = \text{remainder}$$

A remainder may be expressed as a fraction by placing the remainder over the divisor. The quotient is $5\frac{2}{3}$ in the above example.

A remainder may also be expressed as a decimal. To express a remainder as a decimal, add a decimal point in the dividend and add zeros to the right of the decimal point. Continue dividing to the desired places.

Example: The problem is carried out three decimal places and rounded to two places.

$$\begin{array}{r} 5.666 \\ 3\overline{)17.000} \\ -15 \\ \hline 2\,0 \\ -1\,8 \\ \hline 20 \\ -18 \\ \hline 20 \\ -18 \\ \hline 2 \end{array} = 5.67$$

* ***Notice that a decimal point is placed to the right of the 5 in the quotient.***

Divide in the following problems. Express remainders as fractions.

1. $12\overline{)185}$ **2.** $8\overline{)3,321}$ **3.** $6\overline{)5,195}$

Express remainders as decimals. * *There may be situations where a problem is carried out three or more places and the answer rounded to two places. The following problems give practice dividing to four places and then rounding to two places.*

Divide to four decimal places and round to two places.

4. $37\overline{)7,990}$ **5.** $12\overline{)4,622}$ **6.** $45\overline{)3,674}$

Division by Multiples of 10

To divide by multiples of 10, move the decimal point in the dividend one place to the *left* for each zero in the divisor.

Examples: $278 \div 100 = 2.78$
$278 \div 10 = 27.8$
$278 \div 1{,}000 = .278$

Mentally find the quotients in the following problems. Add zeros to the left where needed.

1. $4{,}190 \div 10 =$ __________ **2.** $43 \div 10 =$ __________

3. $2{,}870 \div 100 =$ __________ **4.** $43 \div 100 =$ __________

5. $5{,}628 \div 1{,}000 =$ __________ **6.** $43 \div 1{,}000 =$ __________

Division of Decimals

When the divisor has a decimal, move the decimal point to the *right* until the divisor is a whole number. Then move the decimal point in the dividend to the *right* as many places as you did in the divisor. You may need to add zeros as you move the decimal point.

Example: $5.2\overline{)156} = 52\overline{)1{,}560.} = 52\overline{)1{,}560.} = 52\overline{)1{,}560.}$ (quotient: 30.)

* ***Position the decimal point in the quotient above the decimal point in the dividend before you begin dividing.***

In the following problems, divide to four decimal places and round to two decimal places. * ***Remember to move the decimal point before you begin dividing.***

1. $24.7\overline{).15000}$ **2.** $.46\overline{).210000}$ **3.** $90.2\overline{)36.2}$

4. $6.1\overline{)23.4}$ **5.** $87.63\overline{)6.2}$ **6.** $.0256\overline{).329}$

Division by 1¢ and 10¢

To divide a number by multiples of .01 (1¢), move the decimal point one digit to the *right* for each decimal place in the divisor.

Examples: 378 ÷ .1 = 3,780
378 ÷ .01 = 37,800

Mentally divide in the following problems.

1. 6.43 ÷ .1 = ____________ **2.** 45.9 ÷ .01 = ____________

3. .271 ÷ .001 = ____________ **4.** 8.2 ÷ .1 = ____________

5. 553 ÷ .01 = ____________ **6.** 209.6 ÷ .001 = ____________

Practice

The following problems give practice in division by 10, 100, 1,000, .01, .1, and multiples of 10. Mentally work the problems as quickly as you can. Do not round answers.

Is your calculator turned off?

1. 2,904 ÷ 100 = 29.04
2. 5.3 ÷ .01 = ______
3. .722 ÷ .01 = ______
4. 129 ÷ 30 = ______
5. 800 ÷ 1,000 = ______
6. 19.70 ÷ .01 = ______
7. 1,963 ÷ 10 = ______
8. 2.407 ÷ .1 = ______

Unit Price

Goods are often priced in quantities. To find the **unit price**, divide the group price by the number of units in the group. Any fraction of a cent is counted as a whole. Therefore, when dividing to determine unit price, carry solutions to at least three places and always *round up* to two places. * ***This is one of the few times you don't use the 5/4 rule. The reason is that any fraction of a cent is passed on to the consumer.***

Example: Five records are on sale for $24.92. What would one record cost?

$24.92 ÷ 5 = $4.984 = $4.99

Find the unit price in each problem.

Did you get $1.66?

		Unit Price
1.	Four packages of light bulbs for $6.63	______
2.	Three two-liter bottles of soft drink for $2.67	______
3.	Twenty double-sided $5\frac{1}{4}$" floppy disks for $18.95	______
4.	Six iced tea glasses for $27.50	______
5.	Six nine-volt batteries for $7.79	______

Quantity Pricing

Goods may be priced in groups. The following abbreviations identify groups priced by multiples of 10.

Per **C** = by the 100

Per **M** = by the 1,000

CWT (Hundredweight) = per 100 pounds

Use the shortcut for items priced by C, M, or CWT: Mentally move the decimal point following the procedure for dividing by multiples of 10, then multiply by the selling price.

To find *group price,* divide the total number of units by the quantity of the group. Then multiply by the selling price.

Example: Find the group price of 320 pencils selling at $7.50 per C.

320 ÷ 100 = 3.2 (the number of 100's in 320)
3.2 × $7.50 = $24

* *Do not round the quantity when dividing by multiples of 10.*
Find the group price in each of the following problems.

1. 450 items @ $6 per C = ____________

2. 880 items @ $4 per 100 = ____________

3. 6,200 items @ $2.50 per M = ____________

Try Problem 4 mentally.

4. 578 items @ $10 per 1,000 = ____________

5. 730 lb @ $6.30 per CWT = ____________

6. 1,500 items @ $25 per 1,000 = ____________

Goods may also be priced by a **dozen**, a **ream**, a **gross**, or a **ton**. The quantities and abbreviations are listed below.

doz (dozen) = 12	**ton** = 2,000 pounds
rm (ream) = 500	**gr** (gross) = 144

Remember, to find the group price, you divide the quantity purchased by the number of items in the group and multiply by the group price.

Example: Find the price of 33 items selling at $3.10 per doz.

33 ÷ 12 = 2.75 (dozens purchased)
2.75 (dozens purchased) × $3.10 per doz = $8.525
= $8.53 (total cost)

* *Note that you do not round up to two places until you get the total cost. Carry the quotient to three places before rounding.*

Find the selling price in each of the following problems.

7. 486 @ $15 per doz = ____________

8. 2,450 @ $20 per ton = ____________

9. 1,296 @ $7.75 per gr = ____________

10. 5,700 @ $22.75 per rm = ____________

11. 350 @ $8.50 per gr = ____________

12. 72 @ $5.25 per doz = ____________

Estimating Answers

Are you developing the habit of estimating your answers? Estimating division problems is done by rounding the divisor and dividend to a place value that can be divided mentally.

Example: 6,320 ÷ 198 = 31.9192

rounds to

6,000 ÷ 200 = 30

* ***A misplaced decimal point is the cause of many errors. An estimated answer of 3.19192 is unreasonable because the decimal point is in the wrong position.***

Give the *estimated* answer for each of the following problems.

1. 5,040 ÷ 225 = ____________ **2.** 26 ÷ .48 = ____________

3. 276.16 ÷ 21 = ____________ **4.** 86.3 ÷ 5.9 = ____________

5. 7,860 ÷ 384 = ____________ **6.** 69 ÷ 6.5 = ____________

Now, divide to see if your estimated answers are close to the actual answers. * ***Round to two places.***

Is your calculator turned off?

7. $225\overline{)5{,}040}$ **8.** $.48\overline{)26}$ **9.** $21\overline{)272.16}$

10. $5.9\overline{)86.3}$ **11.** $384\overline{)7,860}$ **12.** $6.5\overline{)69}$

* *Use the estimating method to verify your answers when dividing both mentally and with the calculator.*

Simple Averages

To find a **simple average,** add the numbers in the group and divide the sum of the numbers by the number of items in the group.

Example: $16 + 32 + 18 + 24 + 12 = 102$
$102 \div 5 = 20.4$

Find the average of each of the following groups of numbers. Round each average to two decimal places.

1. 20, 40, 35, 15, 45 __________

2. 200, 250, 240, 220, 245 __________

3. 1,386.24; 1,027.90; 1,165.67; 1,438.42 __________

4. 437, 576, 320, 416 __________

5. 98, 78, 83, 65, 92 __________

Reprinted by permission of UFS, Inc.

CALCULATOR INSTRUCTIONS

Locate the following:

Division Key

Counting Key—not on all calculators

Average Key—not on all calculators

Solve the problem: 1,620 ÷ 45 =

* *Enter the problem into the calculator exactly as it is read: dividend ÷ divisor = quotient*

1. Set the decimal selector on 2.
2. Check your posture.
3. Clear the calculator.
4. Enter 1,620 and operate the division key.
5. Enter 45 and operate the equals key.
   Did you get 36?

Rounding Answers

The more decimal places you work with in arriving at the final answer, the more accurate the answer. If a precise answer is needed, the decimal selector is set on the floating position. The final answer is rounded to the number of places needed.

Calculators differ in the number of decimal places carried out. Some carry out one, two, or three places; others may carry out two, four, or six places; and some carry out the exact number of decimal places up to five or six. In quantity pricing on page 52, you were taught to carry out four places and use the four places to compute the next calculation. The more decimal places you work with in any calculation requiring two or more steps, the more accurate the final answer. Calculators that carry out a maximum of three decimal places may give answers that are one or two cents different than calculators that carry out four or more places.

Example: Items are priced at $2.48 per 12. What is the cost of 29 items?

Decimal Selector on 2:
$2.48 ÷ 12 = $.21 × 29 = $6.09

Decimal Selector on 3:
$2.48 ÷ 12 = $.207 × 29 = $6.003 = $6.00

Decimal Selector on 4:
$2.48 ÷ 12 = $.2067 × 29 = $5.9943 = $5.99

A common business practice is to leave the decimal selector set on 5 or 6 and ONLY round up for the final calculation.

Practice

Find the quotient for each problem below. Round your answers to the nearest thousandth. * *Remember, ending zeros to the right of the decimal point do not change the value and should not be recorded in the answer.*

Is your rounding selector on 5/4 and decimal selector on 3 or 4?

1. 3)177   59
2. 29)21,547
3. 344)57,104
4. 93)63,121
5. 6,181)94,331
6. 35)429

In Problems 7-12, round the quotients to the nearest hundredth.

Did you change your decimal selector to 2?

7. 135)6,075 **8.** 344)1,008 **9.** 216)749

10. 7.34)83.6 **11.** 374)687.65 **12.** 19.5)655.27

Constant Division

When the divisor is the same in a series of division problems, the divisor can be used as a *constant*. The procedure to follow is similar to constant multiplication. * ***Refer to Appendix A for instructions on the constant entry feature.***

Solve the problem: 31 ÷ 25 =
67 ÷ 25 =
98 ÷ 25 =

1. Set the constant mode if necessary.
2. Set the decimal selector on 2.
3. Set the rounding selector on 5/4.
4. Check your posture.
5. Clear the calculator.
6. Enter 31 and operate the division key.
7. Enter 25 and operate the equals key.
   Did you get 1.24?
   Do not clear your calculator!
8. Enter 67 and operate the equals key.
   Did you get 2.68?
9. Enter 98 and operate the equals key.
   Did you get 3.92?

Use constant division to solve these problems.

Is your decimal selector on 2?

1. 679 ÷ 34 = ________
891 ÷ 34 = ________
254 ÷ 34 = ________
79 ÷ 34 = ________

2. 5,274 ÷ 36.5 = ________
4,320 ÷ 36.5 = ________
1,530 ÷ 36.5 = ________
8,040 ÷ 36.5 = ________

3. 525.23 ÷ 12.5 = ________
104.78 ÷ 12.5 = ________
691.72 ÷ 12.5 = ________
436.21 ÷ 12.5 = ________

4. 843 ÷ 2.5 = ________
634 ÷ 2.5 = ________
2,587 ÷ 2.5 = ________
9,066 ÷ 2.5 = ________

5. 1,460 ÷ 84 = ________
1,325 ÷ 84 = ________
1,718 ÷ 84 = ________
1,096 ÷ 84 = ________

6. 3,995 ÷ .49 = ________
286 ÷ .49 = ________
435 ÷ .49 = ________
129 ÷ .49 = ________

Simple Averages

If you have an *average key* (avg.), use it to compute simple averages. Once the sum of a series of numbers is found, operate the average key.

The *counting key* can also be used to compute simple averages. The counting key will tell you how many numbers have been entered into the calculator. To compute the average after the sum of a series of numbers is found, operate the counting key and then divide by the number given.

* *Be sure to read your operating manual for specific instructions.*

Find the average of each of the following groups of numbers. * *Round answers to two places.*

1. 631, 521, 582, 627, 493 ____________

2. 17.9, 18.6, 20.3, 18.4, 19.5, 14.7, 16.7, 18.3 ____________

3. 24, 29, 30, 22, 31, 28, 27 ____________

4. 620.78, 419.22, 600.33, 598.45, 84.60 ____________

5. 2,930; 2,734; 2,645; 2,838; 2,920; 2,756; 2,419 ____________

EVALUATING YOUR SKILLS

You will be timed on the following 20 problems. Your goal is to complete all problems within 10 minutes using your calculator. Your grade will be determined by the number of correct answers. Round quotients to two decimal places.

Division

1. 8,651 ÷ 70 = ____________
2. 86,519 ÷ 34 = ____________
3. 88,076 ÷ 471 = ____________
4. 40,294 ÷ 152 = ____________
5. 62,358 ÷ 403 = ____________
6. 322.2 ÷ .92 = ____________
7. 4.257 ÷ .82 = ____________
8. 630.84 ÷ 72.8 = ____________
9. 227.8 ÷ 67.2 = ____________
10. 82,168 ÷ 42 = ____________

Constant Division

11. 8,260 ÷ 16 = ____________
12. 5,719 ÷ 16 = ____________
13. 6,943 ÷ 16 = ____________
14. 4,803 ÷ 16 = ____________
15. 746 ÷ 12.15 = ____________
16. 256 ÷ 12.15 = ____________
17. 492 ÷ 12.15 = ____________
18. 795 ÷ 12.15 = ____________

Simple Average

19. 250 + 267 + 283 + 259 = ____________ Average ____________
20. 419 + 430 + 429 + 427 = ____________ Average ____________

If you did not get at least 17 correct answers, go back and practice the problems in this unit. Then repeat the problems above.

Grading Scale

Grade	Number of Correct Answers
A	18-20
B	16-17
C	14-15
D	12-13

APPLYING YOUR KNOWLEDGE

Using the information given, calculate the correct answers. * *Remember, unless otherwise instructed, round all answers to two places using the 5/4 position.*

1. Tokuda's Consulting Firm purchased five new computer desks at a total cost of $649.75 and five office chairs at a total cost of $349.75. What did they pay for a computer desk and chair combination? __________

2. A group of employees from a consulting company were going to a seminar in a neighboring town. They needed to travel on a toll road that charged $.75 per car plus $.50 for each additional passenger. How many employees were riding in the car if the total toll cost was $2.75? __________

3. The XYZ Company has a monthly payroll of $13,581.36. To the nearest dollar, what is the average monthly earnings for employees if there are nine employees. __________

4. Find the unit price of each of the following items and the grand total.

Item	Quantity	Unit Price	Total
Footballs	59	__________	$ 646.05
Soccer Balls	36	__________	$ 358.92
Basketballs	68	__________	$1,084.60
Tennis Balls	52 pkg	__________	$ 153.34
		Grand Total	________

5. Jim sold a stereo for $929. He received a down payment of $125. He will receive the balance in 12 equal payments. How much is each payment? __________

6. The Tokyo stock exchange traded a volume of 450 million shares on Monday, 456 million shares on Tuesday, 430 million shares on Wednesday, 420 million shares on Thursday, and 459 million shares on Friday. What was the average daily volume of shares traded?

____

7. Mr. Ferron set aside $149,250 to divide among his employees for Christmas bonuses. If there are 398 employees, how much will each employee receive?

____

8. Katie keeps a daily record of the number of customers who purchase furniture from her and the total amount of sales.

Day	Customer Sales	Total Dollar Sales
Monday	5	$4,250
Tuesday	7	$5,880
Wednesday	4	$1,347
Thursday	1	$640
Friday	6	$4,500

a. What was the average number of customer sales per day?

____

b. What was the average sale per customer?

____

9. The chart below lists the annual salary for the salespeople at Jantz Jewelry Company. Find the monthly salary and weekly salary of each employee. * *Use constant division.*

	Annual Salary	Monthly Salary	Weekly Salary
Alice Clark	$21,320	________	________
Oscar Eller	$21,580	________	________
Tanya Lynn	$18,928	________	________
Stanley Tice	$23,504	________	________
Nancy Vinson	$19,500	________	________

10. The following sale items were advertised in the local paper. Find the savings per unit.

a. 425 grams (15 oz) Beef Ravioli:
Regular price is $1.05 a can.
Sale price is 3 cans for $2.70.

____

b. Size C or D flashlight batteries: Regular price is $1.57 a pkg.
Sale price is 4 pkg for $5.15.

____

c. 4 light bulbs: Regular price is $1.59 per package.
Sale price is 3 packages for $3.95.

____

11. Find the cost of each item given below. Round UP to two places.
* *Do not round until the last calculation is completed. Move your decimal selector to 4.*

a. 150 ft Floor Surfacing Paper @ $35.71 per M

____

b. 125 rolls Joint Tape @ $4.75 per doz

____

c. 1,230 Envelopes @ $9.50 per rm

____

ADVANCED APPLICATIONS

Using the information given, calculate the answers.

1. Sam gets paid $5.25 for assembling each unit. It takes Sam 30 minutes to assemble 1 unit. He assembles 60 units working 6 hours a day. How many days does he work?

____

2. Turf Garden Center bought 336 clay pots from a manufacturer at $21 a dozen. Assuming all pots are sold at $2.50 each, what is the gross profit on the sale of the pots?

____

3. Glen Pollok ordered a shipment of men's leather belts from Italy. When they arrived, they were sized in centimeters. Make a conversion chart for waist size to post above the belts to let the customers know what size belt to purchase.

Waist Size in Centimeters		Waist Size in Inches	Waist Size in Centimeters		Waist Size in Inches
75	=	______	85	=	______
80	=	______	90	=	______
82.5	=	______	95	=	______

* *Divide each unit by 2.5. (Use constant division.)*

4. The Spaulding law firm wants to lower office employee expenses. They presently employ three office personnel at an average yearly cost of $18,750 each. If the company purchased a word processor, only one office person would be needed. The expenses for this person for the year would be: Salary—$15,250; Benefits—$3,500; Information Processing Equipment—$9,431; Workstation—$300; and Supplies—$600. For the one employee and the word processor, determine:

   a. Cost for the year

   ____

   b. Average cost per week (52 weeks = 1 year)

   ____

   c. Average cost per day (1 week = 5 days)

   ____

   d. Average cost per hour (1 day = 8 hours)

   ____

   e. What would be the yearly savings if a word processor were purchased?

   ____

5. Rollin's Assessories had a total inventory of $14,357 on October 1. During the month they purchased additional merchandise that cost $4,250.

   a. What was the amount of inventory during the month?

   ____

   b. Rollin's Assessories took inventory at the end of October and found that the inventory was $7,630. What was the cost of goods sold?

   ____

   c. The total sales for October amounted to $15,929. What was the gross profit?

   (Gross Profit = Sales – Cost of Goods Sold)

   ____

   d. Operating expenses for October amounted to $1,420. What was the net profit for October?

   (Net Profit = Gross Profit – Operating Expenses)

   ____

Wanted
Part-time student to work as trainee in our Accounting Department. Could lead to full-time employment for a career minded individual.

The Hodgepodge Shop

series I A Retail Application

"If you were the boss, would you hire you?"

The Hodgepodge Shop is an exciting and innovative gift shop. All kinds of gifts, fun paper products, cards, stationery items, and accessories contribute to the gala atmosphere. The main store and office is located downtown with three branch stores in shopping malls around the city. The office at the downtown store is looking for a part-time student who will train in the accounting department with the possibility of moving into full-time employment after graduation.

After completing this application, you should be able to:

1. complete a job application form
2. verify sales tickets
3. record information on a Daily Sales Report
4. complete and verify purchase invoices
5. evaluate your personal characteristics and work habits

TERMINOLOGY

The following terminology is used in this application. Become familiar with the meaning of each term so that you understand its usage.

1. **Cash Over**—the cash deposited in the bank is *more than* total cash received from cash sales and paid on account.
2. **Cash Short**—the cash deposited in the bank is *less than* total cash received from cash sales and paid on account.
3. **Daily Sales Report**—summary of sales for one day.
4. **Paid on Account**—money paid to a business to decrease the amount owed by the customer.
5. **Purchase Invoice**—a business paper that shows the items purchased, the quantity, the unit price, and the total price.

6. **Sales Tax**—a tax charged by a state, county, or city on the sales price of consumer goods and/or services.
7. **Sales Ticket**—a form showing the merchandise sold, the quantity, the unit price, the extension, the sales tax, and the total.

JOB APPLICATION FORM

You have decided to apply for the trainee position in the accounting department at The Hodgepodge Shop. An application form has been sent to you to complete. Because the person hired will be working with numbers, neatness and completeness are important and may be used to decide who is hired.

Complete the Application for Employment (Form 1, page 67). Print the information in ink. When you have finished, give the application to your supervisor.

EMPLOYMENT TEST

The Hodgepodge Shop gives each trainee an employment test. This test measures your ability to work with numbers. Since you will use a calculator on the job, you may use your calculator for the Employment Test (Form 2). Give the employment test to your supervisor when you are finished.

ON-THE-JOB-TRAINING

Congratulations! You have been hired as an accounting trainee in our accounting department at The Hodgepodge Shop. You will be doing different jobs throughout your training period so that you will understand how the business office operates. We expect you to do your best at each job assignment. Our company likes to hire trainees as full-time employees after graduation. Your job performance now as a trainee will determine if a full-time position will be offered to you.

In this training session, you have two assignments. Read the job assignments carefully. Follow each step outlined for you. Do not skip any steps. Good Luck! Remember, you are a trainee and should ask your supervisor for help as you need it.

Assignment 1: Sales Tickets

Each time a customer makes a purchase, the salesperson writes the quantity, description, unit price, and total amount of sale on a sales ticket. The extensions are totaled and a 7 percent sales tax is added. The sales tickets are prepared in duplicate. The original is given to the customer, and the copy is sent to the accounting department at the end of the day. A customer's signature is obtained for all charge sales. Accuracy is extremely important because the information is taken from the sales slips to prepare the Daily Sales Report. The Daily Sales Report compiles sales information on the total cash sales and the total charge sales. It compares the actual cash received with what the sales tickets show. In addition, the Daily Sales Report shows the type of merchandise purchased. The Hodgepodge Shop classifies the merchandise into five categories. Each category is indicated by a store stock number recorded on each sales ticket. The stock numbers and classifications are:

XL-1—Paper Products: Plates, napkins, cups, tableclothes

ST-2—Stationery Items: Address books, calendars, photo albums, note pads, invitations, notes, scrapbooks, diaries, memo pads, pencil cups, desk pads, desk accessories

CD-3—Cards, Gift Wrap, and Ribbon

AC-4—Accessories: Jewelry, purses, scarves, belts, sweatshirts, T-shirts, billfolds

GF-5—Gifts: All gift items including photo frames, music boxes, mugs, bells, glassware, silk flowers

Your first responsibility is to verify the daily sales tickets from the previous day's sales and then prepare the Daily Sales Report. Work carefully—accuracy is extremely important.

October 10

1. Remove the Sales Tickets (Forms 3 through 22). Separate the sales tickets on the perforated lines and place them by your calculator.
2. Verify the extension and subtotal on each sales ticket. To verify the sales tax, multiply the subtotal by .07. Add the sales tax to the subtotal to verify the total amount of each sale. If you find an error, draw a line through the incorrect figure and write in the correct figure above the error. Use the corrected figure for calculations.

Did you use accumulative multiplication?

3. Identify each item on the sales tickets by classification number: XL-1 = Paper Products, ST-2 = Stationery Items, CD-3 = Cards-Gift Wrap-Ribbon, AC-4 = Accessories, GF-5 = Gifts. Write the correct stock numbers on the sales tickets.
4. Separate the cash tickets from the charge tickets.
5. Remove the Daily Sales Report (Form 23). In the *cash* tickets, find the total sales of all *paper products*. Verify this total. Record the paper product's sales total in the cash column on the Daily Sales Report.
6. Follow Step 5 to find the total sales for each classification of merchandise. If no merchandise was sold, write —0— on the Daily Sales Report. Be sure to verify your totals.
7. Add the amounts for the five classifications of merchandise and record the sum after Subtotals in the cash column.
8. Find the total tax charged on *cash* sales and verify the total. Write the total on the Daily Sales Report.
9. Repeat Steps 5-8 for *charge* sales using the charge column to record your totals.
10. Find the total daily sales for *cash* sales and then for *charge* sales. Add the Daily Sales Report first and then add the sales tickets. Verify your accuracy by comparing the totals. The charge sales tickets will be given to the accounts receivable person to record in each customer's account. You may file all the sales tickets.
11. On the Daily Sales Report, find the total daily sales by adding the total cash sales and total charge sales. Record this total.
12. Remove the Cash Paid on Account Tickets (Forms 24 through 27). Separate the cash paid on account tickets and place them by your calculator. Total the cash paid on account and verify the total. Record the total on the Daily Sales Report. Since the

cash paid on account tickets are sent to the accounts receivable person, you may file the tickets.

13. Total the Cash Sales and Cash Paid on Account and record the sum as the Total Cash Received from customers. This is the amount of cash that should have been deposited. (If corrections were made on the sales tickets, use the corrected amounts. The amount of an error and/or any errors made when making change will cause the cash deposited to be short or over.) Verify your Daily Sales Report.
14. The amount of cash deposited for the day was $775.55. Record the amount on the Daily Sales Report.
15. The amount of the Daily Sales Report does not agree with the actual amount deposited in the bank because of errors on the sales slips or errors in making change. Indicate the difference on the Daily Sales Report. If the amount is short, place parentheses around it. Submit the Daily Sales Report to your supervisor.

Assignment 2: Purchase Invoices

Your second responsibility is to verify purchase invoices. Purchase invoices are verified before sending them to the accounts payable person for payment.

October 11

Did you use accumulative multiplication?

1. Remove the Purchase Invoices (Forms 28 through 30). Separate the purchase invoices and place them beside your calculator.
2. Verify the extensions. If you find an error, draw a line through the incorrect figure and write in the correct figure. Verify the total of each invoice.
3. Obtain a grand total of all the purchase invoices. Verify this total.
4. Remove the Purchase Authorization (Form 31). Record the total amount purchased and authorize the form by signing it. Submit this form to your supervisor to approve, and then file the invoices.

SELF-EVALUATION

Success in business depends on the right attitudes toward a job and proper work habits. Now that you have completed your first training session, you should evaluate your attitudes and work habits.

1. Remove the Self-Evaluation (Form 32). Rate yourself on each attitude and work habit by checking the appropriate column. Carefully consider each question before ranking yourself.
2. On the lines below, list three areas in which you need to improve.

____

____

____

Discuss with your supervisor how these areas are important in keeping any job. Make an effort during the next few weeks to improve yourself before again working for The Hodgepodge Shop. Remember, good habits are as easily formed as are bad habits.

5. File the self-evaluation form.

Form 1

APPLICATION FOR EMPLOYMENT

AN EQUAL OPPORTUNITY EMPLOYER

PLEASE PRINT WITH BLACK INK OR USE TYPEWRITER

NAME (LAST, FIRST, MIDDLE INITIAL)	SOCIAL SECURITY NUMBER	CURRENT DATE

ADDRESS (NUMBER, STREET, CITY, STATE, ZIP CODE)	HOME PHONE NO.

REACH PHONE NO.	U.S. CITIZEN? YES NO	DATE YOU CAN START

ARE YOU EMPLOYED NOW? IF SO, MAY WE INQUIRE OF YOUR PRESENT EMPLOYER?

TYPE OF WORK DESIRED	REFERRED BY	SALARY DESIRED
		$

IF RELATED TO ANYONE IN OUR EMPLOY, STATE NAME AND POSITION

DO YOU HAVE ANY PHYSICAL CONDITION THAT MAY PREVENT YOU FROM PERFORMING CERTAIN KINDS OF WORK? YES NO IF YES, EXPLAIN

HAVE YOU EVER BEEN CONVICTED OF A FELONY? YES NO IF YES, EXPLAIN

EDUCATION	EDUCATIONAL INSTITUTION	LOCATION (CITY, STATE)	DATES ATTENDED FROM MO.	YR.	TO MO.	YR.	DIPLOMA, DEGREE, OR CREDITS EARNED	CLASS STANDING (CHK QUARTER) 1	2	3	4	MAJOR SUBJECTS STUDIED
	COLLEGE											
	HIGH SCHOOL											
	GRADE SCHOOL						X	X	X	X	X	X
	OTHER											

LIST BELOW THE POSITIONS THAT YOU HAVE HELD (LAST POSITION FIRST)

1. NAME AND ADDRESS OF FIRM	DESCRIBE POSITION RESPONSIBILITIES
NAME OF SUPERVISOR	
EMPLOYED (MO-YR) FROM: TO:	REASON FOR LEAVING

2. NAME AND ADDRESS OF FIRM	DESCRIBE POSITION RESPONSIBILITIES
NAME OF SUPERVISOR	
EMPLOYED (MO-YR) FROM: TO:	REASON FOR LEAVING

3. NAME AND ADDRESS OF FIRM	DESCRIBE POSITION RESPONSIBILITIES
NAME OF SUPERVISOR	
EMPLOYED (MO-YR) FROM: TO:	REASON FOR LEAVING

I UNDERSTAND THAT I SHALL NOT BECOME AN EMPLOYEE UNTIL I HAVE SIGNED AN EMPLOYMENT AGREEMENT WITH THE FINAL APPROVAL OF THE EMPLOYER AND THAT SUCH EMPLOYMENT WILL BE SUBJECT TO VERIFICATION OF PREVIOUS EMPLOYMENT, DATA PROVIDED IN THIS APPLICATION, ANY RELATED DOCUMENTS, OR RESUME. I KNOW THAT A REPORT MAY BE MADE THAT WILL INCLUDE INFORMATION CONCERNING ANY FACTOR THE EMPLOYER MIGHT FIND RELEVANT TO THE POSITION FOR WHICH I AM APPLYING, AND THAT I CAN MAKE A WRITTEN REQUEST FOR ADDITIONAL INFORMATION AS TO THE NATURE AND SCOPE OF THE REPORT IF ONE IS MADE.

____

SIGNATURE OF APPLICANT

EMPLOYMENT TEST

Form 2

Use your calculator to complete the following problems. Recommended time: 15 minutes.

1.

217.04
74.88
(42.46)
573.36
+ (9.15)

2.

729
729
68
354
454
870
+200

3.

813.22
42.91
608.35
S
73.00
956.84
T

4.

65 × 27.66
41 × 20.25
+18 × 37.50

5. 184 × 52 = ______________________

6. 776.2 − 358.56 = ______________________

7. 10.21 × 17 = ______________________

8. 37 ÷ 4 = ______________________

9. 2,843.12 − 2,976.50 = ______________________

10. 12.72 ÷ 6.8 = ______________________

11. Compare the numbers in Column 1 with the numbers in Column 2. Put an X in Column 3 if the figures in the two columns are different.

Column 1	Column 2	Column 3
221.00	221.00	________
678.42	687.43	________
539.27	539.72	________
1,246.50	1,246.50	________
814.73	814.73	________
601.52	610.52	________
92.83	92.83	________
405.30	505.30	________
790.14	790.14	________
356.82	356.83	________

12. How many weeks are in a year? ______________

13. How many months are in a year? ______________

14. How many days are in a year? ______________

15. How many semimonthly pay periods are in a year? ______________

The Hodgepodge Shop Form 3
705 4th Ave. S.
Minneapolis, MN 55415-2232

Cash
CUSTOMER

DATE	SALES NO.	SALESPERSON	CHARGE
10/9/--	005732		

QTY.	STOCK #	DESCRIPTION	UNIT	AMOUNT
2		luncheon plates	3 25	6 50
2		beverage napkins	2 00	4 00
1		7-oz. cups	2 25	2 25
3		cards	2 00	6 00
1		card	1 25	1 25
			SUBTOTAL	20 00
			TAX	1 40
			TOTAL	21 40

CUSTOMER SIGNATURE

The Hodgepodge Shop Form 4
705 4th Ave. S.
Minneapolis, MN 55415-2232

Cash
CUSTOMER

DATE	SALES NO.	SALESPERSON	CHARGE
10/9/--	005733		

QTY.	STOCK #	DESCRIPTION	UNIT	AMOUNT
2		card	1 00	2 00
1		gift enclosure card	30	30
4		gift wrap	1 65	6 60
2		ribbon	2 00	4 00
1		mug	7 00	7 00
1		brass bell	13 75	13 75
			SUBTOTAL	33 65
			TAX	2 36
			TOTAL	36 01

CUSTOMER SIGNATURE

The Hodgepodge Shop Form 5
705 4th Ave. S.
Minneapolis, MN 55415-2232

Pam Onah
CUSTOMER
4010 Pillsbury South
Minneapolis, MN 55409-2386

DATE	SALES NO.	SALESPERSON	CHARGE
10/9/--	005734		✓

QTY.	STOCK #	DESCRIPTION	UNIT	AMOUNT
1		music box	20 00	20 00
1		photo frame	15 00	15 00
2		cards	95	1 90
1		card	1 50	1 50
			SUBTOTAL	38 40
			TAX	2 69
			TOTAL	41 09

Pam Onah
CUSTOMER SIGNATURE

The Hodgepodge Shop Form 6
705 4th Ave. S.
Minneapolis, MN 55415-2232

Cash
CUSTOMER

DATE	SALES NO.	SALESPERSON	CHARGE
10/9/--	005735		

QTY.	STOCK #	DESCRIPTION	UNIT	AMOUNT
3		invitations	2 25	6 75
1		applique belt	9 50	9 50
3		tie dye scarf	4 00	12 00
2		cards	2 50	5 00
			SUBTOTAL	33 75
			TAX	2 36
			TOTAL	36 11

CUSTOMER SIGNATURE

The Hodgepodge Shop
705 4th Ave. S.
Minneapolis, MN 55415-2232

Form 7

CUSTOMER: Cash

DATE	SALES NO.	SALESPERSON	CHARGE
10/9/--	005736		

QTY.	STOCK #	DESCRIPTION	UNIT	AMOUNT
1		man's billfold	21 95	21 95
1		customized T-shirt	12 50	12 50
1		flip photo album	15 00	15 00
			SUBTOTAL	49 45
			TAX	3 46
			TOTAL	52 91

CUSTOMER SIGNATURE

The Hodgepodge Shop
705 4th Ave. S.
Minneapolis, MN 55415-2232

Form 8

CUSTOMER: Judy Lund
1002 Evergreen Ln. N
Minneapolis, MN 55441-0201

DATE	SALES NO.	SALESPERSON	CHARGE
10/9/--	005737		✓

QTY.	STOCK #	DESCRIPTION	UNIT	AMOUNT
3		luncheon napkins	2 75	8 25
4		luncheon plates	3 25	13 00
1		cat calendar	9 50	9 50
1		scrapbook	18 00	18 00
2 pr		earrings	2 50	5 00
2		bracelets	4 00	8 00
			SUBTOTAL	61 75
			TAX	4 32
			TOTAL	66 07

CUSTOMER SIGNATURE: Judy Lund

The Hodgepodge Shop
705 4th Ave. S.
Minneapolis, MN 55415-2232

Form 9

CUSTOMER: Dan McIntosh
5404 Wentworth Ave.
Minneapolis, MN 55419-9451

DATE	SALES NO.	SALESPERSON	CHARGE
10/9/--	005738		✓

QTY.	STOCK #	DESCRIPTION	UNIT	AMOUNT
1		letter tray	56 00	56 00
1		letter opener	8 50	8 50
1		desk pad	19 50	19 50
2		gift wrap	2 75	5 50
3		cards	1 25	3 75
3		ribbon	2 50	7 50
			SUBTOTAL	100 75
			TAX	7 05
			TOTAL	107 80

CUSTOMER SIGNATURE: Dan McIntosh

The Hodgepodge Shop
705 4th Ave. S.
Minneapolis, MN 55415-2232

Form 10

CUSTOMER: Cash

DATE	SALES NO.	SALESPERSON	CHARGE
10/9/--	005739		

QTY.	STOCK #	DESCRIPTION	UNIT	AMOUNT
5		bows	75	3 75
4		gift wrap rolls	2 75	11 00
7		gift wrap	1 65	11 55
1		photo frame	13 00	13 00
6		candles	1 75	10 00
2 pr.		mini barrettes	8 00	16 00
			SUBTOTAL	65 30
			TAX	4 57
			TOTAL	69 87

CUSTOMER SIGNATURE

The Hodgepodge Shop
705 4th Ave. S.
Minneapolis, MN 55415-2232

Form 11

CUSTOMER: Cash

DATE	SALES NO.	SALESPERSON	CHARGE
10/9/--	005740		

QTY.	STOCK #	DESCRIPTION	UNIT	AMOUNT
1		address book	15 00	15 00
1		5-year diary	9 25	9 25
2		note pads	4 50	9 00
3		cards	1 25	3 75
2		cards	1 00	2 00
			SUBTOTAL	39 00
			TAX	2 73
			TOTAL	41 73

CUSTOMER SIGNATURE

The Hodgepodge Shop
705 4th Ave. S.
Minneapolis, MN 55415-2232

Form 12

CUSTOMER: Cash

DATE	SALES NO.	SALESPERSON	CHARGE
10/9/--	005741		

QTY.	STOCK #	DESCRIPTION	UNIT	AMOUNT
1		Peru handbag	40 00	40 00
2		cards	1 25	2 50
1		boxed notes	3 50	3 50
			SUBTOTAL	46 00
			TAX	3 22
			TOTAL	49 22

CUSTOMER SIGNATURE

The Hodgepodge Shop
705 4th Ave. S.
Minneapolis, MN 55415-2232

Form 13

CUSTOMER: Rami Dhanaraj
4200 Freemont Ave. S.
Minneapolis, MN 55420-1719

DATE	SALES NO.	SALESPERSON	CHARGE
10/9/--	005742		✓

QTY.	STOCK #	DESCRIPTION	UNIT	AMOUNT
25		luncheon plates	3 25	81 25
20		luncheon napkins	2 75	55 00
25		9-oz cups	2 25	56 25
25		dessert plates	2 00	50 00
5		silk flower arrangements	25 00	125 00
* Special order				
			SUBTOTAL	367 50
			TAX	25 73
			TOTAL	393 23

Rami Dhanaraj
CUSTOMER SIGNATURE

The Hodgepodge Shop
705 4th Ave. S.
Minneapolis, MN 55415-2232

Form 14

CUSTOMER: Cash

DATE	SALES NO.	SALESPERSON	CHARGE
10/9/--	005743		

QTY.	STOCK #	DESCRIPTION	UNIT	AMOUNT
1		thank you notes	2 25	2 25
1		T-shirt	12 50	12 50
3		barrettes	6 00	18 00
			SUBTOTAL	32 75
			TAX	2 29
			TOTAL	35 04

CUSTOMER SIGNATURE

Form 15

The Hodgepodge Shop
705 4th Ave. S.
Minneapolis, MN 55415-2232

Cash
CUSTOMER

DATE	SALES NO.	SALESPERSON	CHARGE
10/9/--	005744		

QTY.	STOCK #	DESCRIPTION	UNIT	AMOUNT
1		glass bell	11 25	11 25
3		shopping list pads	5 00	15 00
			SUBTOTAL	26 25
			TAX	1 84
			TOTAL	28 09

CUSTOMER SIGNATURE

Form 16

The Hodgepodge Shop
705 4th Ave. S.
Minneapolis, MN 55415-2232

Cash
CUSTOMER

DATE	SALES NO.	SALESPERSON	CHARGE
10/9/--	005745		

QTY.	STOCK #	DESCRIPTION	UNIT	AMOUNT
1		wreath	24 75	24 75
1		photo frame	17 50	17 50
2		cards	1 50	3 00
2		cards	2 00	4 00
1		card	2 50	2 50
			SUBTOTAL	51 75
			TAX	3 62
			TOTAL	55 37

CUSTOMER SIGNATURE

Form 17

The Hodgepodge Shop
705 4th Ave. S.
Minneapolis, MN 55415-2232

Cash
CUSTOMER

DATE	SALES NO.	SALESPERSON	CHARGE
10/9/--	005746		

QTY.	STOCK #	DESCRIPTION	UNIT	AMOUNT
2		ribbon	2 00	4 00
3		gift wrap	1 65	4 95
2		note pads	4 50	9 00
			SUBTOTAL	17 95
			TAX	1 26
			TOTAL	19 21

CUSTOMER SIGNATURE

Form 18

The Hodgepodge Shop
705 4th Ave. S.
Minneapolis, MN 55415-2232

Cash
CUSTOMER

DATE	SALES NO.	SALESPERSON	CHARGE
10/9/--	005747		

QTY.	STOCK #	DESCRIPTION	UNIT	AMOUNT
2		cards	1 00	2 00
			SUBTOTAL	2 00
			TAX	14
			TOTAL	2 14

CUSTOMER SIGNATURE

The Hodgepodge Shop
705 4th Ave. S.
Minneapolis, MN 55415-2232

Form 19

Delores Gonzales
CUSTOMER
225 7th St. N
Minneapolis, MN 55403-5340

DATE	SALES NO.	SALESPERSON	CHARGE
10/9/--	005748		✓

QTY.	STOCK #	DESCRIPTION	UNIT		AMOUNT	
1		music box	22	50	22	50
3		magnets		85	2	55
2		silk plants	125	00	250	00
1		Peru handbag	40	00	40	00
			SUBTOTAL		315	05
			TAX		22	05
			TOTAL		337	10

Delores Gonzales
CUSTOMER SIGNATURE

The Hodgepodge Shop
705 4th Ave. S.
Minneapolis, MN 55415-2232

Form 20

Cash
CUSTOMER

DATE	SALES NO.	SALESPERSON	CHARGE
10/9/--	005749		

QTY.	STOCK #	DESCRIPTION	UNIT		AMOUNT	
1		sports calendar	9	50	9	50
1		pencil cup	7	50	7	50
1		letter holder	13	50	13	50
2		cards	2	00	4	00
1		card	1	25	1	25
			SUBTOTAL		35	75
			TAX		2	50
			TOTAL		38	25

CUSTOMER SIGNATURE

The Hodgepodge Shop
705 4th Ave. S.
Minneapolis, MN 55415-2232

Form 21

Cash
CUSTOMER

DATE	SALES NO.	SALESPERSON	CHARGE
10/9/--	005750		

QTY.	STOCK #	DESCRIPTION	UNIT		AMOUNT	
4		bows		75	3	00
2		gift wrap rolls	2	75	5	50
			SUBTOTAL		8	50
			TAX			60
			TOTAL		9	10

CUSTOMER SIGNATURE

The Hodgepodge Shop
705 4th Ave. S.
Minneapolis, MN 55415-2232

Form 22

Cash
CUSTOMER

DATE	SALES NO.	SALESPERSON	CHARGE
10/9/--	005751		

QTY.	STOCK #	DESCRIPTION	UNIT		AMOUNT	
2		dessert plates	2	00	4	00
2		luncheon napkins	2	75	5	50
1		butterfly pin	4	00	4	00
			SUBTOTAL		13	50
			TAX			95
			TOTAL		14	45

CUSTOMER SIGNATURE

DAILY SALES REPORT

Form 23

DATE ________ / ________ / ________

RECEIPTS FROM SALES

CLASSIFICATION NUMBER	DESCRIPTION	CASH AMOUNT			CHARGE AMOUNT				
XL-1	Paper Products								
ST-2	Stationery Items								
CD-3	Cards, Gift Wrap								
AC-4	Accessories								
GF-5	Gifts								
							*TOTAL DAILY SALES MAY BE USED FOR STATE SALES TAX REPORTS.		
	SUBTOTALS			+			=		
	TOTAL TAX COLLECTED			+			=		
	TOTAL DAILY SALES			+			=		
		+							
	CASH PAID ON ACCOUNT								
	TOTAL CASH RECEIVED								
	CASH DEPOSITED IN BANK								
	CASH OVER OR SHORT								

The Hodgepodge Shop Form 24
705 4th Ave. S.
Minneapolis, MN 55415-2232

CUSTOMER: Mike Moreland
5019 Skyline Dr.
Minneapolis, MN 55436-6907

DATE	SALES NO.	SALESPERSON	CHARGE
10/9/--	005752		

QTY.	STOCK #	DESCRIPTION	UNIT	AMOUNT
		paid on account		125 55
			SUBTOTAL	
			TAX	
			TOTAL	125 55

CUSTOMER SIGNATURE

The Hodgepodge Shop Form 25
705 4th Ave. S.
Minneapolis, MN 55415-2232

CUSTOMER: Van Quinc Ho
5673 Pennsylvania Ave. S.
Minneapolis, MN 55426-0791

DATE	SALES NO.	SALESPERSON	CHARGE
10/9/--	005753		

QTY.	STOCK #	DESCRIPTION	UNIT	AMOUNT
		paid on account		43 70
			SUBTOTAL	
			TAX	
			TOTAL	43 70

CUSTOMER SIGNATURE

The Hodgepodge Shop Form 26
705 4th Ave. S.
Minneapolis, MN 55415-2232

CUSTOMER: Vickie Osborn
6471 14th St. W.
Minneapolis, MN 55426-3211

DATE	SALES NO.	SALESPERSON	CHARGE
10/9/--	005754		

QTY.	STOCK #	DESCRIPTION	UNIT	AMOUNT
		paid on account		70 80
			SUBTOTAL	
			TAX	
			TOTAL	70 80

CUSTOMER SIGNATURE

The Hodgepodge Shop Form 27
705 4th Ave. S.
Minneapolis, MN 55415-2232

CUSTOMER: Amanda Schroeder
4100 Wooddale Ave. S.
Minneapolis, MN 55416-7149

DATE	SALES NO.	SALESPERSON	CHARGE
10/9/--	005755		

QTY.	STOCK #	DESCRIPTION	UNIT	AMOUNT
		paid on account		26 60
			SUBTOTAL	
			TAX	
			TOTAL	26 60

CUSTOMER SIGNATURE

INVOICE

Form 28

ACCESSORIES & MORE
P.O. Box 3271
Boston, MA 02200-6057

INVOICE NO. 6341

DATE October 6, 19--

SOLD TO The Hodgepodge Shop
705 4th Ave. S.
Minneapolis, MN 55415-2232

SALESPERSON	YOUR ORDER NO.	DATE RECEIVED	DATE SHIPPED	SHIPPED BY	TERMS:
JN	K348	9/29/--	10/2/--	UPS	3/15, n/30

QUANTITY	CAT. NO.	DESCRIPTION	UNIT PRICE	EXTENSION
15	7433	Ponytail Barrettes	3.50	52.50
20	6174	Men's Gold ID Bracelet	17.25	345.00
20	6987	Assorted Hair Bows	5.20	104.00
			INVOICE TOTAL $	501.50

INVOICE

Form 29

INVOICE NO. 24973

DATE October 5, 19--

SOLD TO The Hodgepodge Shop
705 4th Ave. S.
Minneapolis, MN 55415-2232

Southern Stationery Supply Co.
700 9th Avenue
New York, NY 10019-4322

SALESPERSON	YOUR ORDER NO.	DATE RECEIVED	DATE SHIPPED	SHIPPED BY	TERMS:
KS	K324	9/27/--	10/04/--	Truck	2/10, n/30

QUANTITY	CAT. NO.	DESCRIPTION	UNIT PRICE	EXTENSION
16	BN-65	Thank you notes	1.50	24.00
12	DY-37	Calendars - Bears	7.00	84.00
12	DY-42	Calendars - Cars	7.00	84.00
12	DY-51	Calendars - Scenic	7.00	84.00
8	WZ-77	Address Books	10.50	84.00
			INVOICE TOTAL $	360.00

INVOICE

Form 30

INVOICE NO. 068019

DATE October 5, 19--

SOLD TO The Hodgepodge Shop
705 4th Ave. S.
Minneapolis, MN 55415-2232

GOODMAN'S GIFT HOUSE
P.O. Box 22211
San Francisco, CA 94100-0385

SALESPERSON	YOUR ORDER NO.	DATE RECEIVED	DATE SHIPPED	SHIPPED BY	TERMS:
DL	K350	9/26/--	9/28/--	Truck	2/15, n/60

QUANTITY	CAT. NO.	DESCRIPTION	UNIT PRICE	EXTENSION
12	32-9055	Four 19-oz Football Glasses	21.95	263.40
12	32-9066	Four 19-oz Daisy Glasses	21.95	263.40
8	32-3824	Set of four Bulgarian Dessert Goblets	32.00	256.00
8	39-3630	Set of six German Glass Bowls	15.00	120.00
8	39-3790	2 qt Chinese Casserole	25.00	200.00
8	39-0803	Set of four Czechoslovakian Footed Compotes	30.00	240.00
			INVOICE TOTAL	$ 1,342.80

PURCHASE AUTHORIZATION

Form 31

The Hodgepodge Shop
705 4th Ave. S.
Minneapolis, MN 55415-2232

Total Amount Purchased ______________________

Authorized by ______________________

____
SUPERVISOR'S INITIALS

SELF-EVALUATION

ATTITUDES	Always	Usually	Seldom	Never
1. Are you cheerful?				
2. Can you control your temper?				
3. Do you do a job you dislike with a good nature and attitude?				
4. Do you avoid gossip?				
5. Do you avoid using profanity?				
6. Can you accept criticism?				
7. Do you accept those whose opinions differ from yours?				
8. Do you believe you can do your work well?				
9. Do you refrain from talking to your friends about personal things during class/work time?				
10. Do you avoid daydreaming?				
11. Are you cooperative and friendly?				
WORK HABITS				
1. Do you get to class/work on time?				
2. Can you do your work without constantly asking questions?				
3. Can you follow instructions without having them repeated?				
4. Is your work neat and complete—the best that you can do?				
5. Are you absent from class/work only when absolutely necessary?				
6. Do you make up excuses if you are tardy or turn in an assignment late?				
7. Do you make it through the day without getting tired?				
8. When you make a mistake do you try to find out why?				
9. Do you use proper grammar?				
10. Do you use your time efficiently?				
11. Do you begin work without being told?				
12. Do you bring all supplies to class/work—pencil, paper, text?				
13. Can you adapt easily to a change of plans?				
14. Do you ever produce more work than is required of you?				
15. Are you dependable?				
Totals				

Total your score giving yourself a 3 for always, 2 for usually, 1 for seldom, and 0 for never. Rate yourself using this chart:

63–78	Outstanding
51–62	Very good
39–50	Improvement needed
Below 39	Poor, much improvement needed

Based on the ratings above, would you give yourself a pay increase or a job promotion? Would you hire you?

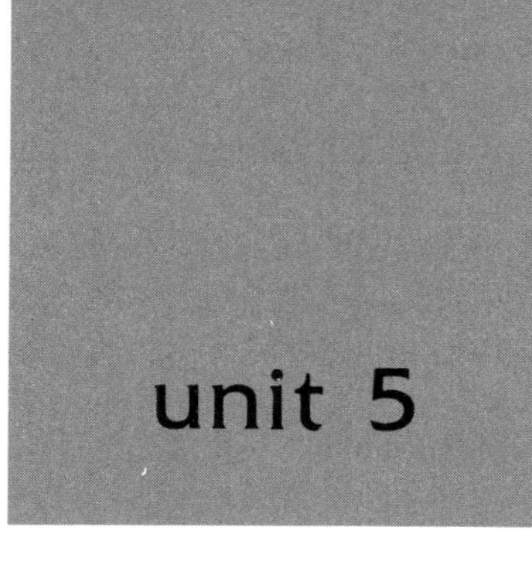
unit 5

Fractions and Decimal Fractions

"No mistake or failure is as bad as to stop and never try again."

When working with invoices, purchase orders, interest, and investment calculations, you need to understand fractions. Once you understand the manual computations, fractions can be converted to decimals. A calculator can then be used for computations.

After studying this unit, you should be able to:

1. identify a proper fraction, an improper fraction, and a mixed number
2. convert a mixed number to an improper fraction and an improper fraction to a mixed number
3. raise a fraction to higher terms and reduce a fraction to lowest terms
4. add, subtract, multiply, and divide fractions and mixed numbers
5. use the cancellation method to simplify multiplication and division of fractions
6. convert fractions to decimals and decimals to fractions
7. use aliquot parts to solve business problems
8. use the calculator to add, subtract, multiply, and divide fractions and mixed numbers
9. use fractions to determine unit pricing
10. solve business problems using fractions and mixed numbers

MATH TERMS AND CONCEPTS

The following terminology is used in this unit. Become familiar with the meaning of each term so that you understand its usage as you develop math skills.

1. **Aliquot Part**—any number that divides another number evenly without a remainder.
2. **Common Denominator**—in a fraction problem, the same denominator in each fraction.
3. **Common Divisor**—a number that will divide evenly into both the numerator and denominator.
4. **Denominator**—in a fraction, the numeral below the line that indicates the number of equal parts into which the whole is divided.

5. **Extending (extension)**—multiplying the quantity by the unit price.
6. **Fraction**—a part of the whole represented by one number placed over another number.
7. **Improper Fraction**—a fraction in which the numerator is equal to or larger than the denominator.
8. **Invert**—in fractions, to reverse the placement of the numerator and the denominator.
9. **Invoice**—a business paper listing merchandise sold.
10. **Least Common Denominator (LCD)**—the smallest denominator evenly divisible by the denominators of a group of fractions.
11. **Mixed Number**—a number containing both a whole number and a fraction.
12. **Numerator**—in a fraction, the numeral above the line that shows how many equal parts of a unit are expressed.
13. **Prime Number**—a number larger than one that can be exactly divided by only itself and one.
14. **Proper Fraction**—a fraction in which the numerator is smaller than the denominator.

Fractions

A **fraction** results when a whole unit is divided into parts. The **numerator** of a fraction represents the part, and the **denominator** represents the whole. For example, four quarters equal $1 (a whole unit). Each quarter represents an equal part of the dollar, or $\frac{1}{4}$. All four quarters equal the whole or $\frac{4}{4}$.

A fraction is another way to state a division problem. Think of the bar that separates the numerator and denominator as a division sign. The fraction $\frac{1}{4}$ can be written as $1 \div 4$, and the division problem $7 \div 24$ can be written as the fraction $\frac{7}{24}$.

Write each of the following division problems as a fraction.

1. $1 \div 2 =$ ________ **2.** $8 \div 25 =$ ________ **3.** $4 \div 7 =$ ________

The two types of fractions are proper fractions and improper fractions. A **proper fraction** is a fraction that has a numerator smaller than the denominator, such as $\frac{7}{8}$. A proper fraction is always less than 1. An **improper fraction** is a fraction equal to or larger than 1. When the numerator is the same as the denominator, such as $\frac{7}{7}$, the fraction equals the whole or 1. When the numerator is larger than the denominator, such as $\frac{8}{7}$, the fraction is greater than 1.

In $\frac{8}{7}$, the denominator indicates that the unit has been divided into seven equal parts. $\frac{8}{7}$ represents one whole unit plus a fractional part of a unit. It is the same as $1\frac{1}{7}$. An improper fraction written as a whole number and a fraction is called a **mixed number**.

To change an improper fraction to a mixed number, find how many whole units are in the numerator by first dividing the numerator by the denominator. Then express any remainder as a proper fraction by placing the remainder over the divisor (denominator).

Example: $\frac{33}{10} = 33 \div 10 = 3\frac{3}{10}$

Change the following improper fractions to mixed numbers or whole numbers.

4. $\frac{16}{16} =$ ________ **5.** $\frac{26}{13} =$ ________ **6.** $\frac{25}{12} =$ ________

7. $\frac{37}{5} =$ ________ **8.** $\frac{8}{7} =$ ________ **9.** $\frac{35}{4} =$ ________

To convert mixed numbers to improper fractions, first multiply the denominator times the whole number. Then add the product to the numerator to determine the equivalent numerator. The denominator in the improper fraction is the same as it is in the mixed number.

Example: Convert $4\frac{3}{5}$ to an improper fraction.

$5 \times 4 = 20$

$20 + 3 = 23$ (numerator)

$4\frac{3}{5} = \frac{23}{5}$ (improper fraction)

Convert the following mixed numbers to improper fractions.

10. $4\frac{2}{5} =$ ________ **11.** $5\frac{3}{4} =$ ________ **12.** $2\frac{2}{3} =$ ________

13. $11\frac{2}{3} =$ ________ **14.** $9\frac{6}{7} =$ ________ **15.** $8\frac{4}{5} =$ ________

Raising to Higher Terms

Remember, when you multiply a number by 1, the value does not change.

The value of a fraction does not change when both the numerator and denominator are multiplied by the same number. Therefore, to raise a fraction to higher terms, multiply the numerator and denominator by the same number.

* ***When the numerator and the denominator of a fraction are the same number, that fraction is equal to 1.***

If you replaced four quarters with twenty nickels, you would still have one dollar or the whole unit; but, instead of $\frac{4}{4}$, you now have $\frac{20}{20}$.

Examples: 1 quarter is the same as 5 nickels, so $\frac{1}{4} = \frac{5}{20}$ $\left(\frac{1}{4} \times \frac{5}{5}\right)$

2 quarters is the same as 10 nickels, so $\frac{2}{4} = \frac{10}{20}$ $\left(\frac{2}{4} \times \frac{5}{5}\right)$

3 quarters is the same as 15 nickels, so $\frac{3}{4} = \frac{15}{20}$ $\left(\frac{3}{4} \times \frac{5}{5}\right)$

Raise the following fractions to higher terms as indicated by the denominator.

1. $\frac{1}{8} = \frac{\quad}{24}$ 2. $\frac{3}{7} = \frac{\quad}{21}$ 3. $\frac{2}{3} = \frac{\quad}{12}$

4. $\frac{3}{4} = \frac{\quad}{48}$ **5.** $\frac{5}{6} = \frac{\quad}{36}$ **6.** $\frac{3}{20} = \frac{\quad}{100}$

Reducing to Lower Terms

Remember, when you divide a number by 1, the value does not change.

A fraction should be reduced to lowest terms. This means that no number *except one* will exactly divide *both* the numerator and denominator. To reduce a fraction to lower terms, divide the numerator and denominator by a number that will exactly divide into both numbers.

* ***When the numerator and denominator of a fraction are the same number, that fraction is equal to 1.***

To reduce $\frac{48}{224}$ you can divide both the numerator and denominator by 2 (which means division by $\frac{2}{2}$) to get $\frac{24}{112}$. Continue to divide by $\frac{2}{2}$ until there is no number that will exactly divide into both the numerator and the denominator.

$$\frac{48}{224} \div \frac{2}{2} = \frac{24}{112}$$

$$\frac{24}{112} \div \frac{2}{2} = \frac{12}{56}$$

$$\frac{12}{56} \div \frac{2}{2} = \frac{6}{28}$$

$$\frac{6}{28} \div \frac{2}{2} = \frac{3}{14}$$

Since there is no number that will exactly divide into both 3 and 14, the fraction is now reduced to lowest terms.

When a fraction is not evenly divisible by 2, then another prime number may be used. **Prime numbers** are 2, 3, 5, 7, 11, 13, 17, 19, 23, 29, etc. The numerator and denominator of the fraction $\frac{5}{15}$ cannot be divided by 2, but both the numerator and denominator can be divided by the prime number 5. The reduced fraction would be $\frac{1}{3}$.

Another method to use in reducing a fraction to lower terms is to find a **common divisor**. Finding the common divisor will save time in reducing fractions.

Example: $\frac{6}{36} \div \frac{6}{6} = \frac{1}{6}$

If a common divisor can not be determined mentally, follow these steps.

Reduce: $\frac{56}{315}$

1. Divide the denominator by the numerator.

$$\begin{array}{r} 5 \\ 56\overline{)315} \\ -280 \\ \hline 35 \end{array}$$

2. Divide the divisor in Step 1 by the remainder.

$$\begin{array}{r} 1 \\ 35\overline{)56} \\ -35 \\ \hline 21 \end{array}$$

3. Continue dividing the divisor by the remainder obtained in each step until there is *no* remainder.

$$\begin{array}{r} 1 \\ 21\overline{)35} \\ -21 \\ \hline 14 \end{array} \qquad \begin{array}{r} 1 \\ 14\overline{)21} \\ -14 \\ \hline 7 \end{array} \qquad \begin{array}{r} 2 \\ 7\overline{)14} \\ -14 \\ \hline \end{array}$$

4. When there is no remainder, the *last divisor* becomes the common divisor. In this example, the common divisor is 7 and is used to reduce $\frac{56}{315}$ to lowest terms:

$$\frac{56}{315} \div \frac{7}{7} = \frac{8}{45}$$

Reduce the following fractions to lowest terms.

1. $\frac{50}{75} =$ ________ **2.** $\frac{35}{56} =$ ________ **3.** $\frac{32}{40} =$ ________

4. $\frac{21}{24} =$ ________ **5.** $\frac{36}{96} =$ ________ **6.** $\frac{15}{45} =$ ________

Addition and Subtraction of Fractions with Common Denominators

When proper fractions have the same denominators (**common denominators**), simply add or subtract the numerators and put the sum or difference over the common denominator.

Examples:

$$\begin{array}{r} \frac{1}{4} \\ +\frac{2}{4} \\ \hline \frac{3}{4} \end{array} \qquad \begin{array}{r} \frac{5}{7} \\ -\frac{3}{7} \\ \hline \frac{2}{7} \end{array}$$

Find the sum or difference in the following problems. Reduce to lowest terms.

1. $\begin{array}{r} \frac{2}{9} \\ +\frac{4}{9} \\ \hline \frac{6}{9} = \frac{2}{3} \end{array}$ **2.** $\begin{array}{r} \frac{7}{8} \\ -\frac{5}{8} \\ \hline \end{array}$ **3.** $\begin{array}{r} \frac{5}{7} \\ +\frac{3}{7} \\ \hline \end{array}$

4. $\frac{8}{11} - \frac{3}{11}$

5. $\frac{1}{3} + \frac{2}{3}$

6. $\frac{3}{4} - \frac{1}{4}$

Addition and Subtraction of Fractions with Unlike Denominators

To add or subtract when the denominators are *not* the same, you first need to find a common denominator. This is done by raising each fraction to higher terms with the **least common denominator (LCD)**. To change each fraction to higher terms using the least common denominator follow these steps.

Solve the problem: $\frac{3}{5} + \frac{1}{3} =$

1. Determine mentally the lowest number the denominators (5 and 3) will divide into with no remainder.

   The least common denominator is 15.

2. Change each fraction to higher terms using the least common denominator: Determine what number to multiply the denominator by to get the LCD. Then multiply the numerator by the same number to form each new fraction.

   $\frac{3}{5} \times \frac{3}{3} = \frac{9}{15}$ $\qquad$ $\frac{1}{3} \times \frac{5}{5} = \frac{5}{15}$

3. Add the numerators and put the sum over the new denominator.

   $\frac{9}{15} + \frac{5}{15} = \frac{14}{15}$

Find the least common denominator, raise each fraction to higher terms, and then add or subtract the fractions. * ***Reduce all fractions to lowest terms.***

1. $\frac{5}{6} = \frac{5}{6}$
   $-\frac{1}{2} = \frac{3}{6}$
   $\frac{2}{6} = \frac{1}{3}$

2. $\frac{6}{7} - \frac{2}{3}$

3. $\frac{3}{4} + \frac{9}{16}$

4. $\frac{2}{3} - \frac{1}{5}$

5. $\frac{1}{2} + \frac{3}{10}$

6. $\frac{4}{5} + \frac{3}{4}$

When you cannot determine a common denominator by sight, use the *LCD Chart*:

Solve the problem: $\frac{7}{12} + \frac{5}{6} + \frac{1}{3} + \frac{7}{9} + \frac{5}{8} =$

1. Write the denominators in a row.

   12 6 3 9 8

2. Divide the denominators by any prime number that will go into more than one denominator.

$$3\overline{)12\ \ 6\ \ 3\ \ 9\ \ 8}$$

3. Bring down the quotients and any number not divisible by the divisor.

$$\begin{array}{r} 3\overline{)12\ \ 6\ \ 3\ \ 9\ \ 8} \\ 4\ \ 2\ \ 1\ \ 3\ \ 8 \end{array}$$

4. Repeat the process until there is no possible divisor.

$$\begin{array}{r} 2\overline{)4\ \ 2\ \ 1\ \ 3\ \ 8} \\ 2\ \ 1\ \ 1\ \ 3\ \ 4 \end{array}$$

$$\begin{array}{r} 2\overline{)2\ \ 1\ \ 1\ \ 3\ \ 4} \\ 1\ \ 1\ \ 1\ \ 3\ \ 2 \end{array}$$

5. Find the product of the divisors and the numbers on the bottom row. The product is the LCD.

$3 \times 2 \times 2 \times 1 \times 1 \times 1 \times 3 \times 2 = 72$ (LCD)

6. Raise each fraction to higher terms using the LCD. Add the fractions.

$$\begin{aligned} \frac{7}{12} &= \frac{42}{72} \\ \frac{5}{6} &= \frac{60}{72} \\ \frac{1}{3} &= \frac{24}{72} \\ \frac{7}{9} &= \frac{56}{72} \\ \frac{5}{8} &= \frac{45}{72} \\ & \quad \frac{227}{72} = 3\tfrac{11}{72} \end{aligned}$$

Use the LCD Chart to find the lowest common denominator. Then add. If the answer is an improper fraction, change it to a mixed number and reduce to lowest terms.

1. $\frac{4}{6}$, $\frac{3}{16}$, $\frac{7}{24}$, $\frac{15}{32}$

2. $\frac{13}{21}$, $\frac{9}{14}$, $\frac{5}{7}$

3. $\frac{4}{15}$, $\frac{1}{9}$, $\frac{3}{5}$, $\frac{2}{3}$

Did you reduce all answers to lowest terms?

Reprinted by permission of UFS, Inc.

Addition of Mixed Numbers

To add mixed numbers, add the *fractions* first. Then add the *whole numbers.* The *sum of the whole numbers* is added to the *sum of the fractions.*

Example:

$$\begin{array}{r} 3\frac{3}{8} \\ +2\frac{15}{16} \\ \hline \end{array}$$

1. Add the fractions (change to fractions with a common denominator):

$$\begin{array}{rcl} \frac{3}{8} & = & \frac{6}{16} \\ +\frac{15}{16} & = & +\frac{15}{16} \\ \hline & & \frac{21}{16} = 1\frac{5}{16} \end{array}$$

2. Add the whole numbers: $3 + 2 = 5$
3. Add the sum of the fractions to the sum of the whole numbers: $5 + 1\frac{5}{16} = 6\frac{5}{16}$.

Find the sums of the mixed numbers. Reduce all fractions to lowest terms.

1. $\begin{array}{r} 2\frac{5}{6} \\ +3\frac{1}{6} \\ \hline \end{array}$

2. $\begin{array}{r} 16\frac{5}{12} \\ +\ 9\frac{1}{8} \\ \hline \end{array}$

3. $\begin{array}{r} 7\frac{1}{4} \\ +2\frac{1}{4} \\ \hline \end{array}$

4. $\begin{array}{r} 4\frac{2}{3} \\ +4\frac{1}{8} \\ \hline \end{array}$

5. $\begin{array}{r} 6\frac{4}{7} \\ +3\frac{1}{3} \\ \hline \end{array}$

6. $\begin{array}{r} 3\frac{8}{9} \\ +7\frac{1}{3} \\ \hline \end{array}$

Subtraction of Mixed Numbers

To subtract mixed numbers, subtract the *fractions* first. Then subtract the *whole numbers.* The *difference of the whole numbers* is added to the *difference of the fractions.*

Example:

$$\begin{array}{r} 6\frac{3}{8} \\ -2\frac{2}{4} \\ \hline \end{array}$$

1. Subtract the fractions.
   a. Change to fractions with a common denominator.

$$\begin{array}{r} \frac{3}{8} = \frac{3}{8} \\ -\frac{2}{4} = \frac{4}{8} \\ \hline \end{array}$$

   b. Borrow a whole unit $(\frac{8}{8})$ from the whole number 6 and add it to the numerator.

$$\begin{array}{r} \overset{5}{\cancel{6}}\frac{3}{8} + \frac{8}{8} = 5\frac{11}{8} \\ -2\frac{4}{8} \quad = 2\frac{4}{8} \\ \hline \end{array}$$

   c. Subtract:

$$\begin{array}{r} \frac{11}{8} \\ -\frac{4}{8} \\ \hline \frac{7}{8} \end{array}$$

2. Subtract the whole numbers: $5 - 2 = 3$. * ***Remember, the 6 became 5 when you borrowed a whole unit in Step 1b.***
3. Add the fraction to the whole number: $3 + \frac{7}{8} = 3\frac{7}{8}$.

Find the difference in each of the following problems. Reduce all fractions to lowest terms.

1. $\begin{array}{r} 12\frac{1}{3} \\ -\ 4\frac{2}{3} \\ \hline \end{array}$

2. $\begin{array}{r} 8\frac{1}{2} \\ -2\frac{1}{3} \\ \hline \end{array}$

3. $\begin{array}{r} 22\frac{2}{3} \\ -16\frac{5}{8} \\ \hline \end{array}$

4. $\begin{array}{r} 7\frac{5}{8} \\ -4\frac{3}{4} \\ \hline \end{array}$

Multiplication of Fractions

To multiply fractions, simply multiply the numerators to get the numerator in the product and multiply the denominators to get the denominator in the product. Then reduce the product to lowest terms.

Example: $\frac{4}{6} \times \frac{1}{3} = \frac{4}{18} = \frac{2}{9}$

Find the products in the following problems. Reduce all fractions to lowest terms.

1. $\frac{3}{4} \times \frac{1}{2} =$ ______________

2. $\frac{3}{8} \times \frac{5}{8} =$ ______________

3. $\frac{3}{2} \times \frac{2}{3} =$ ______________

4. $\frac{5}{9} \times \frac{4}{7} =$ ______________

5. $\frac{1}{2} \times \frac{3}{5} \times \frac{5}{7} =$ ______________

6. $\frac{1}{2} \times \frac{3}{7} \times \frac{2}{3} =$ ______________

To multiply a whole number by a fraction, change the whole number to a fraction by placing it over 1.

Example: $3 \times \frac{2}{5} = \frac{3}{1} \times \frac{2}{5} = \frac{6}{5} = 1\frac{1}{5}$

Find the products in the following problems. Reduce all fractions to lowest terms.

7. $6 \times \frac{3}{7} =$ ____________ **8.** $\frac{3}{5} \times 2 =$ ____________

9. $16 \times \frac{1}{4} =$ ____________ **10.** $\frac{5}{9} \times 5 =$ ____________

11. $15 \times \frac{4}{5} =$ ____________ **12.** $4 \times \frac{3}{8} =$ ____________

Multiplication of Mixed Numbers

To multiply mixed numbers, change the mixed numbers to improper fractions and multiply as fractions.

Example: $1\frac{1}{2} \times 3\frac{1}{5} = \frac{3}{2} \times \frac{16}{5} = \frac{48}{10} = 4\frac{8}{10} = 4\frac{4}{5}$

Find the products in the following problems. Reduce all fractions to lowest terms.

1. $2\frac{3}{4} \times 1\frac{1}{10} =$ ____________ **2.** $1\frac{1}{3} \times 3\frac{2}{5} =$ ____________

3. $3\frac{3}{8} \times 2\frac{2}{3} =$ ____________ **4.** $7\frac{1}{4} \times \frac{4}{10} =$ ____________

Did you reduce your answers and/or change to mixed numbers?

5. $2\frac{1}{4} \times 9\frac{1}{2} =$ ____________ **6.** $9\frac{2}{3} \times 2\frac{3}{4} =$ ____________

Halving a Fraction

If you are halving a fraction, simply double the denominator by multiplying by $\frac{1}{2}$.

Example: Find $\frac{1}{2}$ of $\frac{3}{4}$.

$$\frac{3}{4} \times \frac{1}{2} = \frac{3}{8}$$

Find one half of each fraction.

1. $\frac{1}{6}$ ________ **2.** $\frac{1}{5}$ ________ **3.** $\frac{1}{2}$ ________

Division of Fractions

To divide fractions, **invert** the divisor and multiply.

Example: $\frac{4}{5} \div \frac{1}{3} = \frac{4}{5} \times \frac{3}{1} = \frac{12}{5} = 2\frac{2}{5}$

Divide the fractions in the following problems. Reduce all fractions to lowest terms.

1. $\frac{2}{3} \div \frac{2}{9} =$ ________

2. $\frac{6}{7} \div \frac{5}{6} =$ ________

3. $\frac{5}{8} \div \frac{2}{3} =$ ________

4. $\frac{4}{5} \div \frac{7}{10} =$ ________

5. $\frac{7}{9} \div \frac{3}{9} =$ ________

6. $\frac{5}{6} \div \frac{3}{7} =$ ________

Did you invert the divisor?

Cancellation Method

To speed the division and multiplication of fractions, you can use the *cancellation method.* Certain numbers can be cancelled: like numbers, numbers that divide evenly into another, and numbers divisible by the same factor. Cancelling is like reducing to lowest terms since you divide a numerator and denominator by the same number. Cancel vertically and/or diagonally, but not horizontally. Then multiply or divide as indicated.

Examples:

$\frac{2}{\cancel{5}_1} \times \frac{\cancel{5}^1}{7} = \frac{2}{7}$ *You are dividing the 5's by 5.

$\frac{4}{9} \div \frac{8}{9} = \frac{\cancel{4}^1}{\cancel{9}_1} \times \frac{\cancel{9}^1}{\cancel{8}_2} = \frac{1}{2}$ *You are dividing the 9's by 9 and the 4 and 8 by 4.

Multiply or divide in the following problems, cancelling where possible.

1. $\frac{1}{8} \times \frac{8}{25} \times \frac{5}{7} =$ ________

2. $\frac{4}{15} \times \frac{3}{5} \times \frac{3}{4} =$ ________

3. $\frac{1}{2} \div \frac{7}{10} =$ ________

4. $\frac{7}{8} \div \frac{1}{4} =$ ________

5. $\frac{7}{16} \times \frac{4}{21} \times \frac{4}{5} =$ ______________

6. $\frac{4}{5} \div \frac{3}{15} =$ ______________

Did you reduce to lowest terms by cancelling?

Division of Mixed Numbers

To divide mixed numbers, change the mixed numbers to improper fractions. Then invert the divisor and multiply as fractions.

Example: $3\frac{1}{2} \div 2\frac{1}{5}$

$$\frac{7}{2} \div \frac{11}{5} = \frac{7}{2} \times \frac{5}{11} = \frac{35}{22} = 1\frac{13}{22}$$

Divide the following mixed numbers. Reduce all fractions to lowest terms.

1. $9\frac{1}{2} \div 2\frac{3}{4} =$ ______________

2. $3\frac{5}{6} \div 1\frac{2}{3} =$ ______________

3. $3\frac{1}{4} \div 3\frac{5}{7} =$ ______________

4. $1\frac{4}{5} \div 1\frac{1}{8} =$ ______________

5. $6\frac{1}{2} \div 3\frac{1}{4} =$ ______________

6. $2\frac{1}{3} \div 1\frac{1}{9} =$ ______________

Did you cancel where possible?

Decimal Fractions

To convert a fraction to a decimal, divide the numerator by the denominator.

Example: One quarter $\frac{1}{4}$ of a dollar = $4\overline{)1.00}$ = .25

Change the following fractions to decimals. Carry out to four decimal places.

1. $\frac{1}{3} =$ ________ **2.** $\frac{7}{8} =$ ________ **3.** $\frac{3}{4} =$ ________

Did you write .75 for Problem 3 and not .7500?

4. $\frac{7}{10} =$ ________ **5.** $\frac{4}{5} =$ ________ **6.** $\frac{3}{8} =$ ________

When a mixed number must be converted to a decimal, first change the fraction to a decimal and then add the decimal to the whole number.

Example: Convert $15\frac{1}{4}$ to a decimal.

$1 \div 4 = .25$

$15 + .25 = 15.25$

Conversely, decimals can be converted to fractions. Remember *place value*? The numerator is the decimal. The denominator is a 1 followed by as many zeros as there are digits after the decimal point.

Examples: $.7 = \frac{7}{10}$ $.65 = \frac{65}{100} = \frac{13}{20}$ $3.15 = 3\frac{15}{100} = 3\frac{3}{20}$

Convert the following decimals to fractions or mixed numbers. Reduce all fractions to lowest terms.

7. .75 = ________________ **8.** 2.4 = ________________

9. .625 = ________________ **10.** .35 = ________________

11. .125 = ________________ **12.** 4.25 = ________________

Aliquot Parts

When a number can be divided into the whole number 100 without a remainder, the number is called an **aliquot part**. Aliquot parts make it possible to do many calculations mentally. In some calculations, using aliquot parts (fractional equivalents of decimals) gives a more accurate answer than solving with decimals.

Example: A wholesaler lists parts at $12\frac{1}{2}$ cents each. $12\frac{1}{2}$ is an aliquot part because $100 \div 12\frac{1}{2} = 8$. Each part is $\frac{1}{8}$ of the whole dollar. If a retailer purchases 32 items at $12\frac{1}{2}$¢ each, see how easy it is to solve: $32 \times \frac{1}{8} = \$4$.

The following chart shows aliquot parts of a dollar expressed in cents, as fractions, and as decimal equivalents. Use aliquot parts to simplify business calculations.

Employees who use aliquot parts frequently will memorize them.

ALIQUOT PART CHART

$25¢ = \frac{1}{4} = .25$	$12\frac{1}{2}¢ = \frac{1}{8} = .125$
$50¢ = \frac{1}{2} = .50$	$37\frac{1}{2}¢ = \frac{3}{8} = .375$
$75¢ = \frac{3}{4} = .75$	$62\frac{1}{2}¢ = \frac{5}{8} = .625$
	$87\frac{1}{2}¢ = \frac{7}{8} = .875$
$33\frac{1}{3}¢ = \frac{1}{3} = .3333$	$10¢ = \frac{1}{10} = .10$
$66\frac{2}{3}¢ = \frac{2}{3} = .6667$	$5¢ = \frac{1}{20} = .05$
	$4¢ = \frac{1}{25} = .04$
$16\frac{2}{3}¢ = \frac{1}{6} = .1667$	$20¢ = \frac{1}{5} = .20$
$83\frac{1}{3}¢ = \frac{5}{6} = .8333$	$40¢ = \frac{2}{5} = .40$
	$60¢ = \frac{3}{5} = .60$
	$80¢ = \frac{4}{5} = .80$

Example: When extending an invoice, "think":

32 items @ $12\frac{1}{2}$¢ is $32 \times \frac{1}{8} = \$\ 4$

48 items @ $16\frac{2}{3}$¢ is $48 \times \frac{1}{6} = \ 8$

16 items @ 25¢ is $16 \times \frac{1}{4} = \ 4$

Total $16

Using the aliquot part chart above, find the cost in each of the following problems.

Is your calculator off?

1. 240 items @ 25¢ = ________

2. 550 items @ 80¢ = ________

Did you multiply by $\frac{1}{4}$ in in Problem 1?

3. 200 items @ 5¢ = ________ 4. 560 items @ $62\frac{1}{2}$¢ = ______

5. 320 items @ $37\frac{1}{2}$¢ = ______ 6. 1,000 items @ 50¢ = ______

Practice

Solve the following problems. Reduce all answers to lowest terms.

1. $\frac{1}{4} + \frac{3}{8} =$ ________________
2. $\frac{2}{3} + \frac{7}{8} =$ ________________
3. $\frac{3}{5} - \frac{1}{2} =$ ________________
4. $\frac{11}{12} - \frac{1}{3} =$ ________________
5. $\frac{2}{9} \times \frac{3}{4} =$ ________________
6. $\frac{42}{75} \times \frac{15}{49} =$ ________________
7. $\frac{3}{5} \div \frac{2}{3} =$ ________________
8. $\frac{7}{8} \div \frac{3}{4} =$ ________________

CALCULATOR INSTRUCTIONS

Fractions may need to be converted to decimals to perform calculator operations. * ***A conversion table is often used. Study the table of Decimal Equivalents in Appendix C.***

Addition of Fractions

To add fractions on a calculator, first convert the fractions to decimals by division and accumulate in the grand total or memory register. When adding mixed numbers, accumulate the sum of the whole numbers in the grand total register also. * ***Remember, when possible you should convert mentally and not have to divide on your calculator!***

Solve the problem:

$$\begin{array}{r} 33\frac{1}{7} \\ 67\frac{2}{5} \\ +\ 189\frac{5}{7} \\ \hline \end{array}$$

1. Set the decimal selector to round to four places.
2. Set the calculator to accumulate.
3. Check your posture.
4. Clear your calculator.
5. Convert the fraction $\frac{1}{7}$ to a decimal by division and accumulate in GT or memory.
   Did you get .1429?
6. Repeat Step 5 for the fractions $\frac{2}{5}$ and $\frac{5}{7}$, accumulating in GT or memory.
   Did you get .4 and .7143?
7. Add the whole numbers, 33 + 67 + 189, and accumulate in the grand total or memory.
   Did you get 289?
8. Operate the GT total or memory total.
   Did you get 290.2572?

Find each sum. Round answers to four decimal places.

Is your rounding selector on 5/4 and decimal selector on 4? (If your calculator has only three decimal places, set your decimal selector on FL and manually round to four places.)

1. $\begin{array}{r} 24\frac{1}{2} \\ 55\frac{1}{3} \\ 267\frac{4}{7} \\ \hline \end{array}$ **2.** $\begin{array}{r} 2\frac{4}{9} \\ 57\frac{2}{5} \\ 4\frac{1}{8} \\ 69\frac{3}{8} \\ \hline \end{array}$ **3.** $\begin{array}{r} 148\frac{9}{10} \\ 39\frac{3}{4} \\ 207\frac{5}{7} \\ 65\frac{4}{5} \\ \hline \end{array}$ **4.** $\begin{array}{r} 50\frac{1}{6} \\ 87\frac{2}{3} \\ 7\frac{2}{3} \\ 7\frac{5}{8} \\ \hline \end{array}$ **5.** $\begin{array}{r} 57\frac{2}{3} \\ 290\frac{1}{5} \\ 400\frac{8}{9} \\ 86\frac{5}{6} \\ \hline \end{array}$

Subtraction of Fractions

As in addition of fractions using a calculator, convert the fractions to decimals to subtract. Use GT with the GT equal minus key or memory minus to shorten the subtraction process.

Solve the problem:

$$\begin{array}{r} 765\frac{3}{4} \\ -\ 22\frac{4}{5} \\ \hline \end{array}$$

1. Set the decimal selector to round to four places.
2. Set the calculator to accumulate.
3. Check your posture.
4. Clear your calculator.
5. Convert $\frac{3}{4}$ to a decimal by division and add to the memory.
   Did you get .75?
6. Convert $\frac{4}{5}$ to a decimal by division and subtract from the memory.
   Did you get .8?
7. Subtract the whole numbers, 765 – 22, and accumulate in the GT or memory.
   Did you get 743?
8. Operate the GT total or memory total.
   Did you get 742.95?

Find each difference. Round answers to four decimal places.

1. $35\frac{4}{5} - 22\frac{1}{5}$ 2. $760\frac{3}{7} - 144\frac{5}{8}$ 3. $8{,}512\frac{5}{9} - 2{,}416\frac{1}{2}$ 4. $1{,}029\frac{1}{3} - 847\frac{5}{6}$ 5. $7{,}496\frac{2}{9} - 875\frac{3}{7}$

Multiplication of Fractions

Multiplication of fractions can generally be done in one continuous operation on a calculator. Solve the problem: $\frac{4}{5} \times \frac{2}{3} =$

1. Set the decimal selector to round to four places.
2. Check your posture.
3. Clear your calculator.
4. Enter 4 and operate the division key.
5. Enter 5 and operate the equals key and the multiplication key.
6. Enter 2 and operate the division key.
7. Enter 3 and operate the equals key.
   Did you get .5333?

Solve each problem. Round answers to four decimal places.

1. $\frac{2}{39} \times \frac{4}{5} =$ ________ 2. $\frac{9}{13} \times \frac{4}{7} =$ ________

3. $\frac{2}{17} \times \frac{5}{13} =$ ________ 4. $\frac{2}{9} \times \frac{6}{11} =$ ________

5. $\frac{19}{22} \times \frac{13}{18} =$ ________ 6. $\frac{2}{3} \times \frac{5}{7} =$ ________

Division of Fractions

Remember, to divide fractions you must first invert the divisor and then multiply.

Example: $\frac{7}{8} \div \frac{4}{5} = \frac{7}{8} \times \frac{5}{4} = \frac{35}{32} = 1.0938$

The fractions *can* be converted to decimals and then divided.

Example: $\frac{7}{8} \div \frac{4}{5} = .875 \div .8 = 1.0938$

Solve these problems. Carry calculations to four places and round answers to two decimal places.

1. $\frac{3}{19} \div \frac{4}{13} =$ ________ 2. $\frac{1}{6} \div \frac{5}{9} =$ ________

3. $\frac{1}{2} \div \frac{3}{4} =$ ________ 4. $\frac{12}{17} \div \frac{23}{27} =$ ________

5. $\frac{8}{9} \div 3 =$ ________ 6. $\frac{7}{8} \div \frac{2}{5} =$ ________

Pricing

See Unit 4, page 51, to find unit price.

When comparing prices to determine the best buy, it is sometimes necessary to find the cost of a larger unit. Follow the same procedure used to find unit price to find the cost of a larger unit. Once a single unit price is known, multiply by the whole unit by following *one* of two methods.

1. Determine the unit price and multiply by the total quantity.

OR

2. Divide the cost of the known units by the fraction of the two quantities.

Do you know there are 16 ounces in 1 pound?

Example: If a 12-ounce package of rice costs 60¢, what would a 1-pound package cost?

1. $\frac{\$.60}{12} = \$.05$ (cost per ounce = unit price)

$\$.05 \times 16 = \$.80$ per pound

* *Cancellation will simplify the procedure.*

$\frac{\overset{.05}{\cancel{\$.60}}}{\underset{1}{\cancel{12}}} \times 16 = \$.80$ per pound

2. $\$.60 \div \frac{12}{16} = \frac{\overset{.20}{\cancel{\$.60}}}{1} \times \frac{\overset{4}{\cancel{16}}}{\underset{\underset{1}{\cancel{3}}}{\cancel{12}}} = \$.80$ per pound

Remember, you round up when calculating unit price.

Find the price per pound to the nearest cent. Set your decimal selector on 3 or 4 and round up to two places.

1. 6 oz for 48¢ = ________ **2.** 12 oz for 97¢ = ________

3. 20 oz for $1.40 = ________ **4.** 32 oz for $1.57 = ________

5. 5 oz for 95¢ = ________ **6.** 10 oz for 80¢ = ________

EVALUATING YOUR SKILLS

You will be timed on the following 20 problems. Your goal is to complete all of the problems within 10 minutes using your calculator. Your grade will be determined by the number of correct answers. All answers must be in decimal form. * *Set your decimal selector on 4.*

1. $\frac{1}{6} + \frac{1}{4}$

2. $\frac{1}{3} + \frac{5}{8} + \frac{3}{10}$

3. $\frac{2}{5} + \frac{1}{3} + \frac{2}{9}$

4. $15\frac{2}{3} + 44\frac{1}{7} + 13\frac{1}{2}$

5. $19\frac{5}{7} + 10\frac{3}{8} + 25\frac{5}{6}$

6. $\begin{array}{r} \frac{11}{12} \\ -\frac{3}{4} \\ \hline \end{array}$ **7.** $\begin{array}{r} \frac{4}{5} \\ -\frac{1}{3} \\ \hline \end{array}$ **8.** $\begin{array}{r} \frac{9}{10} \\ -\frac{3}{5} \\ \hline \end{array}$ 9. $\begin{array}{r} 12\frac{7}{10} \\ -\ 8\frac{8}{11} \\ \hline \end{array}$ 10. $\begin{array}{r} 34\frac{7}{12} \\ -26\frac{2}{7} \\ \hline \end{array}$

11. $\frac{3}{7} \times \frac{3}{8} =$ ______________ **12.** $\frac{2}{5} \times \frac{6}{7} =$ ______________

13. $3\frac{1}{2} \times 7\frac{1}{6} =$ ______________ **14.** $15\frac{3}{4} \times 4\frac{3}{7} =$ ______________

15. $20\frac{1}{5} \times 8\frac{4}{7} =$ ______________ **16.** $\frac{7}{8} \div \frac{1}{9} =$ ______________

17. $\frac{3}{8} \div \frac{2}{5} =$ ______________ **18.** $\frac{2}{3} \div \frac{5}{8} =$ ______________

19. $\frac{8}{9} \div 3 =$ ______________ **20.** $5 \div 1\frac{2}{3} =$ ______________

If you did not reach your goal, practice the problems on pages 106 and 107 using your calculator. Then repeat the problems above.

Grading Scale

Grade	Number of Correct Answers
A	18-20
B	16-17
C	14-15
D	12-13

APPLYING YOUR KNOWLEDGE

Using the information given in the problems, calculate the answers. Reduce all fractions to lowest terms. * ***Round answers to two places using the 5/4 rule unless otherwise instructed.***

1. Jake received a shipment of 36 light bulbs. Twelve of the bulbs were broken.

a. Express the number of bulbs broken as a fraction.

____

b. Express the number of bulbs Jake could sell as a fraction.

____

2. Jenny needs to buy 18 inches of material. The material is sold by the yard. What fraction of a yard does she need to buy? (Hint: 1 yard = 36 inches.)

How many cm is this?

3. A store's total sales for one day were $4,500. Of that total, $1,500 were credit charge sales, $550 were layaway sales, and the remainder were cash sales. What fraction of total sales were credit charge sales?

4. An item is sold in two different sizes—$11\frac{1}{5}$ oz and $4\frac{1}{4}$ oz. How many more ounces are in the $11\frac{1}{5}$ oz size than in the $4\frac{1}{4}$ oz size?
   a. Fraction

   b. Decimal

5. In Problem 4, which size costs less per ounce, the $11\frac{1}{5}$ oz size selling for $1.63 or the $4\frac{1}{4}$ oz size selling for $ .80? * *Set decimal selector on 4.*

6. Find the average weight of four gold beads if they weigh $33\frac{4}{5}$ grams, $45\frac{1}{3}$ grams, $41\frac{3}{8}$ grams, and $37\frac{1}{4}$ grams. * *Set decimal selector on 4.*

7. For each of the following, determine the cost per pound. Then underline the item that costs less per pound. Round up to two places.
   a. Item A—14 oz pkg for 66¢

      Item B—1 lb pkg for 87¢

   b. Item G—9 oz for 82¢

      Item H—1 lb 3 oz for $1.02

c. Item C—3 lb pkg for $2.70

Item D—56 oz pkg for $2.80

8. Hightower designs and manufactures Indian jewelry. He began selling his work in several European jewelry stores. Last month he calculated that $\frac{1}{8}$ of all sales came from international sales. If sales totaled $122,576, what was the dollar sales on the international market?

9. An independent mail service charges $12\frac{1}{2}$¢ per magazine. If a magazine publisher sends 133,000 issues a month to 43 states, what is its mailing expense per month?

ADVANCED APPLICATIONS

Calculate the answers for the following problems.

1. A pharmacist charges $4.00 for 70 grams of Brand A medicine. The same medication labeled Brand B costs $2.20 an ounce. Which brand is the better buy?

2. David and Brian each earn $26.70 a day. At the end of five days, David put $\frac{2}{7}$ of his earnings in the bank and Brian put $\frac{1}{3}$ of his earnings in the bank.
   a. Who saved the most?

   b. How much more was saved?

3. The Hopper Health Food Store sells a product in $\frac{4}{5}$-pound packages. Because of higher costs, they are reducing the amount in the packages by $\frac{1}{8}$. What will be the size of the new package in pounds?

4. $7\frac{1}{6} \times 5\frac{1}{12} \times 2\frac{7}{9}$

a. Set your decimal selector on 3. Change each mixed number to a decimal rounded to three decimal places and multiply. ____________

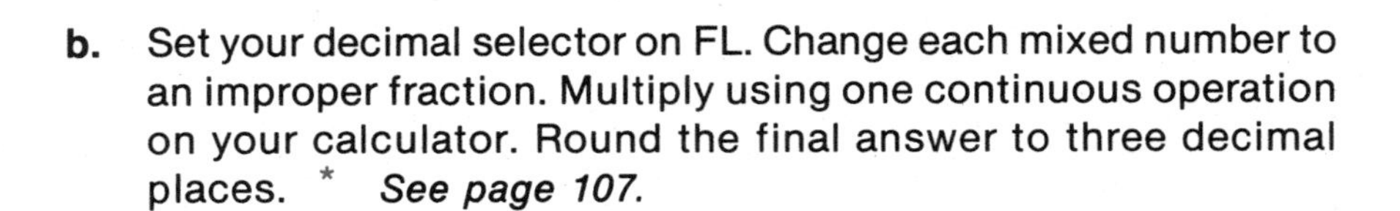

b. Set your decimal selector on FL. Change each mixed number to an improper fraction. Multiply using one continuous operation on your calculator. Round the final answer to three decimal places. * *See page 107.* ____________

c. What is the difference between the product in Part a and the product in Part b? ____________

* *Answer may vary according to number of decimal places used and the number of places carried on the FL position.*

5. Michael Parrott is vice-president of sales for Women's Accessories, Inc. He sold $\frac{2}{3}$ of his cosmetics stock in two hours. How much of the stock could he sell in $\frac{3}{4}$ of an hour? ____________

6. The Good Fruit Farm bought $5\frac{3}{4}$ cases of oranges. It sold $\frac{3}{5}$ the first day, $\frac{1}{8}$ the second day, and $\frac{1}{5}$ the third day. How many cases remained at the end of the third day? * *Set decimal selector on 4.* ____________

unit 6 Multioperations

"Everyone is self made,
but only the successful
want to admit it."

Business problems usually require more than one mathematical process. Combining addition, subtraction, multiplication, and division in a problem can often be done in a continuous process on the calculator.

After studying this unit, you should be able to:

1. solve multioperation equations
2. find a missing item in a series
3. find a weighted average
4. solve multioperation equations using the calculator
5. perform negative multiplication using the calculator
6. solve business math problems using multioperations

MATH TERMS AND CONCEPTS

The following terminology is used in this unit. Become familiar with the meaning of each term so that you understand its usage as you develop math skills.

1. **Multioperations**—combining addition, subtraction, multiplication, and/or division.
2. **Negative Multiplication**—combining multiplication with subtraction.
3. **Weighted Average**—computing the average when the quantity of any unit is more than one.

Multioperations Process

Many business problems require moving from one math operation to another. You should study a problem carefully to determine which process you should do first. In an equation, the portion inside parentheses should be solved first. When there are no parentheses, do the division and multiplication first followed by any addition and/or subtraction—always computing from left to right.

Follow the indicated operations to solve the following problems.

1. $(6 \div 2) + (9 \times 2) - 1 =$ _____ **2.** $(7 + 3) \times 5 + 13 =$ ________

3. $3 + (4 \div 2) - 2 =$ _________ **4.** $3 + 7 - (1 \times 8) + 4 =$ ______

Finding a Missing Item in a Series

When an average is given, but one item in the series is missing, the procedure for finding the missing item is:

1. Multiply the given average by the total number of items in the series.
2. Add the items in the series.
3. Subtract the sum found in Step 2 from the product found in Step 1.

Example: Jason wants an 80 average in his business math class. His scores on the first four tests were 74, 88, 83, and 70. What score must he make on the fifth test to have an 80 average for all five tests?

$80 \times 5 = 400$

$74 + 88 + 83 + 70 = 315$

$400 - 315 = 85$

Jason must score at least 85 on the fifth test to have an 80 average.

Find the missing item in the series for each of the following problems.

Is your calculator turned off!

1. Average wanted of five numbers is 97. Four given numbers are 82, 78, 67, and 90.

____

2. Average wanted of six numbers is 270. Five given numbers are 250, 225, 220, 254, and 190.

____

3. Average wanted of four numbers is 50. Three given numbers are 28, 60, and 35.

____

4. Average wanted of seven numbers is 130. Six given numbers are 100, 120, 125, 140, 130, and 122.

____

Finding a Weighted Average

To find the average if there is more than one unit of an item:

1. Multiply each item by the number of units.
2. Divide the sum of the products by the sum of the units.

Example: Mary's Dress Shop sold 15 dresses at $32.99, 18 dresses at $35.99, and 12 dresses at $29.99. Mary, the shop's owner, wanted to calculate the average price of all dresses sold.

15 × $32.99 =	$ 494.85
18 × $35.99 =	647.82
12 × $29.99 =	359.88
45	$1,502.55

$1,502.55 ÷ 45 = $33.39 (average price of dresses sold)

Find the weighted average for each problem below.

1.

22 @ $14.95 = ________

31 @ $15.66 = ________

25 @ $18.90 = ________

_____ ________

Average ________

2.

312 @ $.75 = ________

370 @ $.60 = ________

389 @ $.72 = ________

_____ ________

Average ________

3.

69 @ $30.00 = ________

65 @ $44.80 = ________

73 @ $39.85 = ________

68 @ $39.93 = ________

_____ ________

Average ________

4.

46 @ $8.95 = ________

57 @ $10.12 = ________

36 @ $15.15 = ________

60 @ $7.42 = ________

_____ ________

Average ________

CALCULATOR INSTRUCTIONS

You can often move from one operation to another on the calculator without having to reenter a previous answer. This simplifies the problem-solving process. Using *memory* or *accumulative* features on your calculator will also save time. Consult your operating manual as you progress through this section.

Addition and Subtraction Followed by Multiplication or Division

Solve the problem: (25 + 16 − 10) × 4.2 =

1. Set the decimal selector on 2.
2. Clear your calculator.
3. Enter 25 and add.
4. Enter 16 and add.
5. Enter 10 and subtract.
6. Obtain a total. * ***Subtotal on some models.***
   Did you get 31?
7. Operate the multiplication key.
8. Enter 4.2 and operate the equals key.
   Did you get 130.2?

* ***Some models will allow you to omit Step 6. Learn the capabilities of your calculator.***

Follow the same procedure for addition and subtraction followed by division.

Solve the problem: $(25 + 16 - 10) \div 4.2 =$

1. Set the decimal selector on 2.
2. Clear your calculator.
3. Enter 25 and add.
4. Enter 16 and add.
5. Enter 10 and subtract.
6. Obtain a difference. (Subtotal on some models).
   Did you get 31?
7. Operate the division key.
8. Enter 4.2 and operate the equals key.
   Did you get 7.38?

* ***Some models will allow you to omit Step 6. Learn the capabilities of your calculator.***

The answers are given in the following problems. Work each problem until you obtain the correct answer.

1. $(9 + 32 - 6) \times 14 = 490$

2. $(9.5 + 29.87 - 13.5) \div 12 = 2.16$

3. $(45 - 17 + 66) \times 58 = 5{,}452$

4. $(76 - 55 + 13.3) \div 4.6 = 7.46$

Complete the following problems. Do not reenter a total to get the answer.

5. $(22 + 43 - 7) \times 6 =$ ________

6. $(574 + 613) \div 363 =$ ________

7. $(66 - 27 + 13) \times 30 =$ ________

8. $(374 + 384) \div 68 =$ ________

9. $(52 - 40.62 + 8.50) \times 18 =$ ________

10. $(857 - 372) \div 14 =$ ________

Multiplication or Division Followed by Addition or Subtraction

Solve the problem: $(120 \times 5) + 12 =$

1. Set the decimal selector on 2.
2. Clear your calculator.
3. Enter 120 and operate the multiplication key.
4. Enter 5 and operate the equals key.
   Did you get 600?
5. Operate the plus key to reenter 600.
6. Enter 12 and add.
7. Obtain a total.
   Did you get 612?

Follow the same procedure for division followed by subtraction.

Solve the problem: $(120 \div 5) - 12 =$

1. Set the decimal selector on 2.
2. Clear your calculator.
3. Enter 120 and operate the division key.
4. Enter 5 and operate the equals key.
   Did you get 24?
5. Operate the plus key to reenter 24.
6. Enter 12 and subtract.
7. Obtain a total.
   Did you get 12?

The answers are given in the following problems. Work each problem until you obtain the correct answer.

1. $(34 \times 6) + 13 = 217$
2. $(200 \div 4) + 18 = 68$
3. $(12 \times 3) - 8 = 28$
4. $(243 \div 3) - 12 = 69$

Complete the following problems.

5. $(30 \times 8) + 15 =$ ______
6. $(3{,}241 \div 820) + 254 =$ ______
7. $(27 \times 6) - 39 =$ ______
8. $(4{,}468 \div 271) - 13.5 =$ ______
9. $(7.1 \times 7) - 17 =$ ______
10. $(27{,}045 \div 779) + 415 =$ ______

Division Followed by Multiplication and Multiplication Followed by Division

The floating position should be used when doing continuous multiplication and division calculations. Mentally round the final answer to the number of decimal places needed. This should give a more accurate answer.

Solve the problem: $(450 \div 50) \times 2 =$

1. Set the decimal selector on F or FL.
2. Clear your calculator.
3. Enter 450 and operate the division key.
4. Enter 50.
5. Operate the equals key.
6. Operate the multiplication key.
7. Enter 2 and operate the equals key.
   Did you get 18?

Follow the same procedure for multiplication followed by division.
Solve the problem: $(450 \times 50) \div 2 =$

1. Set the decimal selector on F or FL.
2. Clear your calculator.
3. Enter 450 and operate the multiplication key.
4. Enter 50 and operate the equals key.
   Did you get 22,500?
5. Operate the division key.
6. Enter 2 and operate the equals key.
   Did you get 11,250?

The answers are given in the following problems. Work each problem until you obtain the correct answer.

Setting the decimal selector on FL will give a more exact final answer than rounding to two places and then completing the operation.

1. $(631 \div 33) \times 40 = 764.85$
2. $(26 \times 4) \div 8 = 13$
3. $(432.1 \div 16) \times 7.56 = 204.17$
4. $(305 \times 13) \div 21.4 = 185.28$

Complete the problems below.

*5. $(612 \div 61) \times 8.2 =$ ______

6. $(77.3 \times 18.3) \div 64.5 =$ ______

*7. $(80.2 \div 7.3) \times 21.62 =$ ______

8. $(188 \times 1.43) \div 14 =$ ______

Did you work each problem in one continuous operation on the calculator?

*9. $(335 \div 5.7) \times 70 =$ ______

10. $(2{,}800 \times 79) \div 350 =$ ______

Practice

Notice in Problem 8 that multiplication follows division. You may want to move your decimal selector to FL for a more accurate answer.

Combination practice on multioperations. ***Set decimal selector on 2.***

1. (104 + 41 − 25) × 17 = ______ **2.** (25.49 + 98.17) ÷ 34.4 = ______

3. (87.74 − 26.62) ÷ 8.4 = ______ **4.** (23 × 26) − 215 = ______

5. (57 × 91) + 72 = ______ **6.** (3,247 ÷ 67) + 29 = ______

7. (437 × 67) ÷ 214 = ______ ***8.** (7,414 ÷ 576) × 12 = ______

9. (51 × 6) ÷ 10 = ______ **10.** (3,541 ÷ 28) − 52 = ______

Negative Multiplication

Solve the problem: (51 × 5.5) − (51 × 4.25) =

1. Set the decimal selector on 2.
2. Clear your calculator.
3. Enter 51 and operate the multiplication key.
4. Enter 5.5 and do *one* of the following:
   a. Grand total models: operate the equals plus key.
   b. Memory models: operate the equals key and/or memory plus key.
   c. Printing calculators: operate the equals key and plus key.
   * ***A few models will not subtract from grand total. Complete each multiplication problem individually and then subtract.***
5. Enter 51 and operate the multiplication key.
6. Enter 4.25 and do *one* of the following:
   a. Grand total models: operate the equal minus key.
   b. Memory models and computer calculator: operate the equals key and/or memory minus key.
   c. Printing calculators: operate the equals key and minus key.
7. Take a total from grand total, memory, or total key.
   Did you get 63.75?

Work these problems.

1. (13 × 15) − (8 × 20)______

2. (50.8 × 74.7) − (34.6 × 81) = ______

3. (8.7 × 77) − (1.32 × 87) = ______

4. (87.2 × 59) − (120 × 36) = ______

Missing Item

The memory feature on most calculators can be used to shorten the problem-solving process when you are solving for a missing item in a series.

Solve the problem: Average wanted of five numbers is 80. Given numbers are 74, 88, 83, and 70. Find the missing item.

1. Set the decimal selector on 0.
2. Clear your calculator.
3. Multiply 80 × 5 and enter the product into memory plus or operate the equals plus key.

4. Add 74 + 88 + 83 + 70 and enter the sum into memory minus or operate your minus key twice followed by the total key to enter it into GT as a negative number.
5. Operate the memory total or grand total key to obtain the difference.

   Did you get 85?

Find the missing item in the series for each of the following problems.

1. Average wanted of five numbers is 25. Four given numbers are 17, 19, 27, and 21.

____

2. Average wanted of three numbers is 13. Two given numbers are 12 and 17.

____

3. Average wanted of seven numbers is 47. Six given numbers are 74, 45, 11, 30, 41, and 32.

____

4. Average wanted of five numbers is 75. Four given numbers are 77, 92, 86, and 71.

____

EVALUATING YOUR SKILLS

You will be timed on the following problems. Your goal is to complete all the problems within 15 minutes. Your grade will be determined by the number of correct answers. Round final answer to two places.

1. (9.34 + 4.22 − 6.2) × 21 = ______________

2. (68.62 + 79.61 − 83.43) ÷ 8.5 = ______________

3. (215 ÷ 10.2) + 340 = ______________

4. (75.6 × 33.7) + 509 = ______________

5. (52.6 × 35.1) ÷ 52 = ______________

6. (15,379 ÷ 356) × 8.3 = ______________

7. (62 × 13.3) − (12 × 5.6) = ______________

8. (123.5 + 78 − 40.3) ÷ 7.3 = ______________

9. (24,530 ÷ 377) − 49 = ______________

10. (12.56 × 29.2) − 97 = ______________

11. (6,804 ÷ 73) + 216 = ______________

12. (74,004 ÷ 172) × 39 = ______________

13. $(21.78 \times 34.41) \div 59 =$ ______

14. $(18{,}130.21 \div 170) - 242 =$ ______

15. $(389 \times 4.7) + 1{,}763 =$ ______

16. $(67.11 \times 60.05) - (34.2 \times 37) =$ ______

17. $(88.64 + 47.53) \div 21 =$ ______

18. $(341 - 123.7) \times 13 =$ ______

If you did not reach your goal, go back and practice the problems on pages 116-118. Then repeat the test before continuing.

Grading Scale

Grade	Number of Correct Answers
A	17-18
B	15-16
C	13-14
D	12

APPLYING YOUR KNOWLEDGE

Work the following problems using the information given. Round answers to two places.

1. Tools & Hardware, Inc. sold the following electrical supplies: 27 plastic guards at 12¢ each, 19 wire caps at 23¢ each, and a bundle of wire for $27.88. When the cashier calculated the cost on the cash register, she got a total charge before sales tax of $45.49.
   a. Are her calculations logical? * *Estimate the answer.* ______

   b. What should be the total charge? ______

2. A manufacturer of computer disk drives wants to complete an average of 105 drives per day. The plant is open five days a week. The plant completed 89 drives on Monday, 97 drives on Tuesday, 111 drives on Wednesday, and 103 drives on Thursday. How many disk drives must be completed on Friday to average 105 drives a day? ______

3. John Austin, vice-president of Tittle Drilling Co., travels around the state inspecting company projects. He uses his own car but is reimbursed 27¢ per mile. During the last year, Mr. Austin drove 19,742 miles.
   a. How much was Mr. Austin reimbursed for the year? ______________

   b. What was the average cost to the company per month? ______________

   c. When he travels in Europe, Mr. Austin is reimbursed 14¢ per kilometer. How much will he be reimbursed if he travels 19,742 miles? ______________

4. A sporting goods store keeps $75 in change in its cash register drawer. During one day $15 was paid out for delivery services and a cash refund was given of $17.71. At the end of the day, the register showed sales receipts of $679.38. There was $781.09 in the cash drawer.
   a. How much should have been in the drawer? ______________

   b. Is the drawer short or over? ______________

   c. By how much? ______________

5. Complete the invoice below. Remember aliquot parts!

	Quantity	Description	Unit Price	Amount
a.	18 cans	Tomatoes	$.33$\frac{1}{3}$	______
b.	32 cans	Olives	.62$\frac{1}{2}$	______
c.	24 cans	Potatoes	.37$\frac{1}{2}$	______
d.	28 cans	Apples	.25	______
e.	40 cans	Peas	.60	______
	Total			______

6. A salesperson wanted to average $2,400 in sales for a four-week period. The first three weeks the salesperson sold $1,900; $2,900; and $2,100. What must the salesperson sell the fourth week to average $2,400? ______________

7. The Plumbing Shop made the following purchases of gasoline for the service truck: 16 gal @ $1.01, 17 gal @ $1.21, 14 gal @ $1.05, and 15 gal @ $1.11. What was the average cost per gallon?

____

8. The local YMCA held a special fund-raising event. They sold 17 record albums at $3.98, 5 cassette tapes at $7.98, 14 record albums at $4.95, and 9 record albums at $6.45. If three of the $3.98 albums were returned, what was the average selling price?

____

9. The Sharp Clothing Store's expenses for the last quarter are given below.
   a. Find the monthly totals and quarterly totals. * *Use grand total.*

	October	November	December	Quarterly Totals
Rent	$1,000.00	$1,000.00	$1,000.00	______
Utilities	755.00	835.69	742.75	______
Insurance	250.00	250.00	250.00	______
Office Supplies	67.34	89.23	52.75	______
Salaries	950.00	930.00	1,040.00	______
Monthly Totals	______	______	______	______

   b. What is the average expense for each item?
      Rent ______________
      Utilities ______________
      Insurance ______________
      Office Supplies ______________
      Salaries ______________
   c. What is the average monthly expense for each month?
      October ______________
      November ______________
      December ______________

Did you use the constant feature?

   d. What is the total average expense per month for the first quarter?

____

ADVANCED APPLICATIONS

Calculate the answers to the following problems.

1. On December 31, Sperling Glass and Mirror Co. had an inventory of $150,920. On January 31, the inventory was $98,432. Goods purchased in January totaled $56,000.

   a. Find the cost of goods sold in January.

   $$\left(\begin{matrix}\text{Beginning}\\\text{Inventory}\end{matrix} + \text{Purchases} - \begin{matrix}\text{Ending}\\\text{Inventory}\end{matrix} = \text{Cost of Goods Sold}\right)$$ ____________

   b. If sales were $176,598, find the gross profit.

   (Sales – Cost of Goods Sold = Gross Profit) ____________

   c. If total expenses were $15,764, find the net profit.

   (Net Profit = Gross Profit – Expenses) ____________

2. A flower shop purchased 16 dozen roses at $28 a dozen. Five of the roses were wilted and had to be thrown away. Five dozen roses were sold at $36.75 a dozen. The remaining roses were sold at $7.50 each. What was the store's gross profit?

   * *Can be worked in one continuous operation.*

   ____

3. If an executive earning $38,000 per year spends an average of 3.2 hours a day in meetings, how much of the yearly salary is earned participating in meetings? (252 working days per year)

   * *Set decimal selector on 4.*

   ____

4. A 660-milliliter bottle of chemicals sells for $2.30. A similar brand sells for $2.50 for 20 fluid ounces.

   a. Which size is the better buy? ____________

   b. How much per ounce is saved? ____________

unit 7 Decimals and Percents

"Getting along with others depends about 98% on your own behavior."

Businesses use decimals and percents every day. Percents are used to show comparison of figures, rate of inflation, taxes, salary, price, and profit. Understanding decimals and percents will allow you to work many business math computations with ease. Before you can work effectively with percents, you should thoroughly understand decimals and fractions. Then it will become easy to move from decimals to percents and from percents to decimals.

After studying this unit, you should be able to:

1. effectively use decimals and fractions
2. use aliquot parts in converting decimals, fractions, and percents
3. convert decimals and fractions to percents and percents to decimals and fractions
4. convert a fractional percent to a decimal
5. find a percent of a number
6. compute sales tax
7. use the calculator to find the percent of a number
8. solve business math problems using decimals, fractions, and percents

MATH TERMS AND CONCEPTS

The following terminology is used in this unit. Become familiar with the meaning of each term so that you understand its usage as you develop math skills.

1. **Fractional Percent**—a portion of 1 percent.
2. **Percent**—comparison of any number to 100, often represented by %.
3. **Sales Tax Table**—a listing of the tax amounts to be applied to sale prices.

You should be able to change fractions to decimals to percents and back with ease. Let's review several concepts.

* ***Remember to watch for errors in decimal placement!***

Rounding Decimals

The total number of digits in a decimal may not be needed. Business situations require different decimal settings. You should be able to round to tenths, hundredths, or thousandths with ease.

In the following problems, round to one, two, and three decimal places. (For a review of rounding, see pages 32-33.)

	Tenth	Hundredth	Thousandth
1. .4217 =	______	______	______
2. 2.80749 =	______	______	______
3. .9631 =	______	______	______
4. 13.0991 =	______	______	______
5. 6.2433 =	______	______	______
6. .0898 =	______	______	______
7. 72.4226 =	______	______	______
8. 4.9467 =	______	______	______
9. 6.8729 =	______	______	______
10. 6.0249 =	______	______	______

Dividing and Multiplying by Multiples of 10 and .1, .01, and .001

When a number is *divided by multiples of 10,* you simply move the decimal point one digit to the *left* for each zero in the divisor. However, if you *divide by .1, .01, or .001,* move the decimal point one digit to the *right* for each decimal place in the divisor.

When a number is *multiplied by multiples of 10,* move the decimal point one digit to the *right* for each zero in the multiplier. However, if you *multiply by .1, .01, or .001,* move the decimal point one digit to the *left* for each decimal place in the multiplier.

Follow the indicated operation to solve each of the following problems. Do not round the answers.

Turn your calculator off.

1. 7.4 ÷ 100 = ______ **2.** .081 ÷ .01 = ______

3. .23 ÷ 10 = ______ **4.** 9.5 × 100 = ______

5. 6.27 × 10 = ______ **6.** 3.78 ÷ .001 = ______

7. .37 × .001 = ______ **8.** 6.74 × 1,000 = ______

9. 3.55 × .01 = ______ **10.** 210 × .01 = ______

Did you move the decimal point in the right direction?

11. 4.32 ÷ .1 = ______ **12.** 62 ÷ 1,000 = ______

* ***For a review of multiplication and division by multiples of 10, see pages 31, 32, 49, and 50.***

Converting Fractions to Decimals

To change a fraction to a decimal, divide the numerator by the denominator. To convert a mixed number to a decimal, convert the fraction to a decimal and add it to the whole number.

Convert the following numbers to decimals. Round to the nearest thousandth.

1. $\frac{7}{8}$ = ____________________ **2.** $8\frac{1}{3}$ = ____________________

3. $3\frac{1}{9}$ = ____________________ **4.** $9\frac{1}{2}$ = ____________________

5. $4\frac{7}{10}$ = ____________________ **6.** $19\frac{2}{3}$ = ____________________

7. $\frac{3}{7}$ = ____________________ **8.** $\frac{11}{12}$ = ____________________

Aliquot Parts

Using aliquot parts will simplify converting fractions to decimals. See how many of the following you can convert to decimal or fractional equivalents by memory. Round each decimal answer to three places.

* *If you need help, the equivalency chart is on page 104.*

1. $\frac{1}{4}$ = ____________________ **2.** .833 = ____________________

3. .75 = ____________________ **4.** $\frac{1}{8}$ = ____________________

5. $\frac{1}{3}$ = ____________________ **6.** $\frac{2}{5}$ = ____________________

7. .667 = ____________________ **8.** .125 = ____________________

Percents

Percent means hundredths—per (by) cent (hundredth). It is a comparison with a hundred and can mean "per hundred," "by the hundred," or "out of a hundred." The hundred is called the base or whole.

If you had 100 cassette tapes and sold 10 tapes, you sold 10 out of a hundred or 10 percent of the tapes. You are actually dividing the whole into 100 parts and comparing what you sold to what you started with. This comparison can be expressed in several ways:

Fraction: $\frac{10}{100}$ (the denominator is always 100 or the whole)
$\frac{10}{100} = \frac{1}{10}$ of the tapes were sold.
* *Always reduce a fraction to lowest terms.*

Decimal: .1 were sold

Percent: 10% were sold

Percent is another way to write a decimal and a fraction. The decimal .1 and the fraction $\frac{1}{10}$ equal 10%. The process of changing a decimal to a percent or a percent to a decimal is a mental one that you should be able to accomplish without the aid of a calculator. The decimal equivalent of the percent is used to solve business math problems.

Changing a Decimal to a Percent

A decimal is changed to a percent by moving the decimal point two places to the right and adding the percent sign (%). (You are multiplying the decimal by 100.)

Examples: $.75 = .75 \times 100 = 75\%$

$.672 = .672 \times 100 = 67.2\%$

$.6 = .6 \times 100 = 60\%$ * *You must add a zero in order to move the decimal point two places.*

Change the following decimals to percents.

Did you multiply each decimal by 100, or did you just move the decimal point two places to the right?

1. .44 = ________ **2.** .07 = ________ **3.** .9 = ________

4. .52 = ________ **5.** .366 = ________ **6.** .25 = ________

7. 2.00 = ________ **8.** .08 = ________ **9.** 1.67 = ________

Changing a Fraction to a Percent

A fraction is changed to a percent by first converting the fraction to a decimal. The decimal is then converted to a percent.

Examples: $\frac{1}{4} = .25 = 25\%$

$\frac{2}{7} = .286 = 28.6\%$

$\frac{1}{8} = .125 = 12.5\%$

Change the following fractions and mixed numbers to percents. Carry the percent out to the nearest tenth of a percent.

Remember, tenth means one decimal place.

	Decimal	Percent		Decimal	Percent
1. $\frac{3}{4}$ =	________ =	________	**2.** $\frac{3}{8}$ =	________ =	________
3. $\frac{1}{2}$ =	________ =	________	**4.** $\frac{1}{5}$ =	________ =	________
5. $\frac{4}{7}$ =	________ =	________	**6.** $4\frac{2}{9}$ =	________ =	________
7. $\frac{3}{10}$ =	________ =	________	**8.** $3\frac{1}{3}$ =	________ =	________
9. $\frac{4}{5}$ =	________ =	________	**10.** $\frac{10}{11}$ =	________ =	________

Did you use aliquot parts?

Changing a Percent to a Decimal

To change a percent to a decimal, move the decimal point two places to the left and drop the percent sign. (You are dividing the percent by 100).

Examples: $14\% = 14 \div 100 = .14$

$6\% = 6 \div 100 = .06$

$25.3\% = 25.3 \div 100 = .253$

$30\% = 30 \div 100 = .3$

Change the following percents to decimals.

Did you divide by 100, or did you just move the decimal point two places to the left?

1. 15% = ________ **2.** .8% = ________ **3.** 24.5% = ________

4. 63% = ________ **5.** 95% = ________ **6.** 22.8% = ________

7. 200% = ________ **8.** 3.6% = ________ **9.** 89.1% = ________

Changing a Percent to a Fraction

To change a percent to a fraction, convert the percent to a decimal. Then convert the decimal to a fraction. Reduce the fraction to lowest terms.

Remember, the number of decimal places indicates the denominator—10, 100, 1,000, etc. Reduce the fraction to lowest terms.

Examples: $24.5\% = .245 = \frac{245}{1,000} = \frac{49}{200}$
$20.5\% = .205 = \frac{205}{1,000} = \frac{41}{200}$

Change the following percents to decimals and then to fractions. Reduce the fractions to lowest terms.

		Decimal	Fraction
1.	7%	= ________	= ________
2.	90%	= ________	= ________
3.	87.5%	= ________	= ________
4.	2%	= ________	= ________
5.	15%	= ________	= ________
6.	60%	= ________	= ________
7.	3.5%	= ________	= ________
8.	12.5%	= ________	= ________
9.	75%	= ________	= ________
10.	80%	= ________	= ________

Did you use aliquot parts?

Changing a Fractional Percent to a Decimal

A **fractional percent** is a portion of 1 percent. It is changed to a decimal by converting the percent to a decimal percent and then changing the decimal percent to a decimal.

Examples: $\frac{1}{2}\% = .5\% = .005$
$\frac{1}{5}\% = .2\% = .002$
$\frac{1}{8}\% = .125\% = .00125$

Change the following fractional percents to decimal percents and then to decimals. Round to four decimal places.

		Decimal %	Decimal
1.	$\frac{1}{4}\%$	= ____________	= ____________
2.	$\frac{5}{10}\%$	= ____________	= ____________
3.	$\frac{2}{5}\%$	= ____________	= ____________
4.	$\frac{2}{7}\%$	= ____________	= ____________
5.	$\frac{1}{3}\%$	= ____________	= ____________

		Decimal %		Decimal
6. $\frac{1}{12}\%$	=	____________	=	____________
7. $\frac{5}{6}\%$	=	____________	=	____________
8. $\frac{8}{9}\%$	=	____________	=	____________
9. $\frac{5}{8}\%$	=	____________	=	____________
10. $\frac{2}{3}\%$	=	____________	=	____________

Did you use aliquot parts?

HERMAN®

"I got 6 percent in math. Is that good or bad?"

HERMAN COPYRIGHT 1987 UNIVERSAL PRESS SYNDICATE

Finding the Percent of a Number

A **percent** expresses a relationship between two numbers. When we say 5%, we mean 5% of some number—the whole. The percent is an amount equal to a part of the whole (100%). The relationship in problem-solving situations may be expressed in *one* of these ways:

1. 5% of 227 equals what amount?
2. Find 5% of 227.
3. What is 5% of 227?

In each way, you are finding the part of the whole represented by 5%.

Examples: 5% of 100 = .05 × 100 = 5 (*of* means to multiply)

5% of 227 = a part of the whole = .05 × 227 = 11.35

In each of the following problems, find the part of the whole represented by the percent. * ***Round answers to two places.***

1. 25% of 250 = ____________ **2.** 9% of 39 = ____________

3. 15% of 33.49 = ____________ **4.** 24% of 680 = ____________

5. Find $7\frac{1}{4}\%$ of 180 ____________ **6.** 4.5% of 20 = ____________

Did you change the percent to a decimal before multiplying?

7. Find 11% of 43 ____________ **8.** What is 36% of 126? ______

9. What is 7% of 65? ______ **10.** 10% of 400 = ____________

Sales Tax

Most states and many cities and counties charge a *sales tax* on merchandise that is sold to consumers. The tax is collected at the place of purchase and then sent to tax agencies. *Sales tax* is a percent that is a part of the whole. It is figured by multiplying the price by the sales tax rate. The tax is then added to the purchase price to obtain the total sales.

Example: $14.95 price; 6% tax rate
$14.95 × 6% = $.897 = $.90 (tax)
$14.95 + $.90 = $15.85 (total sales)

Tax tables are often kept near cash registers to eliminate the need to compute the tax. The tax table below is used for 5%, $6\frac{1}{4}$%, and $7\frac{1}{4}$% tax rates. If tax is figured on even dollars, the table is not needed.

5%		$6\frac{1}{4}$%		$7\frac{1}{4}$%	
AMOUNT	TAX	AMOUNT	TAX	AMOUNT	TAX
$ .01 — .09	No Tax	$ .01 — .07	No Tax	$ .01 — .06	No Tax
.10 — .29	.01	.08 — .23	.01	.07 — .20	.01
.30 — .49	.02	.24 — .39	.02	.21 — .34	.02
.50 — .69	.03	.40 — .55	.03	.35 — .48	.03
.70 — .89	.04	.56 — .71	.04	.49 — .62	.04
.90 — 1.09	.05	.72 — .87	.05	.63 — .75	.05
1.10 — 1.29	.06	.88 — 1.03	.06	.76 — .89	.06
1.30 — 1.49	.07	1.04 — 1.19	.07	.90 — 1.03	.07

Example: If a sales price is $24.73 and the sales tax rate is $6\frac{1}{4}$%, what is the total tax charged?

1. Use the $6\frac{1}{4}$% table to find the tax on $.73.
   $.72 — .87: $.05
2. Figure the tax on the whole dollars.
   $24 × .0625 = $1.50
3. Add the two taxes together to find the total sales tax.
   $.05 + $1.50 = $1.55

Using the tax tables, find the tax on the following sales.

	Sales Price	5%	$6\frac{1}{4}$%	$7\frac{1}{4}$%
1.	$22.14			
2.	$5.22			
3.	$18.43			
4.	$10.02			
5.	$37.77			

Locate the following:

Percent Key—The percent does not have to be changed to a decimal before entering it into the calculator when the percent key is used. (If your calculator does not have a percent key, change all percents to their decimal equivalents.)

* ***Often percent answers are wrong because of a misplaced decimal. Estimate your answer!***

Solve the problem: 20% of 80 =

1. Set the decimal selector on 2.
2. Clear your calculator.
3. Enter 80 and operate the multiplication key.
4. Enter 20 and operate the percent key OR enter .20 and operate the equals key.
   Did you get 16?

Using your calculator, find the percents.

1.	35% of 500 = ________	**2.**	20% of 600 = ________
3.	16% of 375 = ________	**4.**	12.5% of 386 = ________
5.	5.25% of 88 = ________	**6.**	33% of 478 = ________
7.	43% of 876 = ________	**8.**	16% of 674 = ________
9.	8.25% of 254 = ________	**10.**	82% of 1,278 = ________

Some calculators have an add-on percent feature. This feature adds part of the whole (the percent) to the whole.

Locate the Percent Plus Key—not on all calculators

Solve the problem: Find the sales price of a $23.95 item if sales tax is 4%.

1. Enter 23.95 and operate the multiplication key.
2. Enter 4, operate the percent key and then operate the plus key, OR enter 4 and operate the percent plus key. Then the percent will be added to the whole and the calculation is ended.
   Did you get $24.91?

Find the amount of the sales tax and the total sales price for the following.

	Purchase Price	Tax Rate	Sales Tax	Total Sales Price
1.	$58.90	4%	________	________
2.	$45.90	3%	________	________
3.	$.86	6%	________	________
4.	$432.70	7%	________	________
5.	$900.76	$5\frac{1}{4}$%	________	________
6.	$1,850.00	$6\frac{1}{2}$%	________	________
7.	$576.52	$3\frac{1}{4}$%	________	________
8.	$450.43	$4\frac{3}{4}$%	________	________
9.	$286.98	7%	________	________
10.	$177.34	5%	________	________

EVALUATING YOUR SKILLS

You will be timed for 20 minutes on your ability to use aliquot parts to convert decimals, fractions, and percents. One of the three is given. Fill in the two missing items. Carry the percents to the nearest tenth of a percent as a decimal or use the fractional form.

1.	______	=	______	=	______
2.	______	=	$\frac{1}{4}$	=	______
3.	.875	=	______	=	______
4.	.667	=	______	=	______
5.	______	=	______	=	$16\frac{2}{3}\%$
6.	______	=	______	=	5%
7.	______	=	$\frac{1}{2}$	=	______
8.	______	=	$\frac{3}{5}$	=	______
9.	.333	=	______	=	______
10.	.625	=	______	=	______
11.	______	=	$\frac{5}{6}$	=	______
12.	______	=	______	=	30%
13.	______	=	______	=	$12\frac{1}{2}\%$
14.	______	=	$\frac{3}{8}$	=	______
15.	______	=	______	=	10%
16.	.4	=	______	=	______
17.	______	=	$\frac{1}{5}$	=	______
18.	______	=	______	=	4%
19.	______	=	______	=	80%
20.	______	=	$\frac{1}{200}$	=	______

Grading Scale

Grade	Number of Correct Answers
A	35-40
B	30-34
C	25-29
D	20-24

APPLYING YOUR KNOWLEDGE

Using the information given, calculate the correct answers. * *Set decimal selector on 2.*

1. AB Construction Co. purchased a piece of equipment at a cost of $28,430. They had to pay 7% of the purchase price as a down payment. What was the amount of the down payment?

   ____

2. The Chong law firm made $137,769 profit in 1990. Three partners each received $33\frac{1}{3}$% of the profit. What amount did each receive?
   * *Use the fraction aliquot part.*

   ____

3. Determine the amount of The Food Mart's sales tax to be sent to the tax commission if total sales for the month were $57,890, including $4,655 of nontaxable sales. City tax is 2% and state tax is 4.5%
   a. Amount sent to city

   ____

   b. Amount sent to state

   ____

4. The Studio Record Shop rents store space in a shopping center. They must pay 11% of their gross profit per year for rent. If their gross profit is $97,660, how much rent do they owe?

   ____

5. The Jones Company had 18% of its workers absent on Monday due to an outbreak of the flu. How many workers were absent if there are 650 total workers?

   ____

6. The owner of a local grocery store purchased 15 cases of blueberries sold in liter boxes. Each case contained 6 liter boxes. After opening all the cases, the owner discovered about 30% were spoiled and had to be thrown away.
   a. How many liter boxes had to be thrown away?

   ____

   b. Express the total number of spoiled boxes as a fraction.

   ____

7. Luis is the owner of a local confectionery shop. He sold $\frac{3}{8}$ of a case of candy. What percent of the case did he sell?

   ____

8. Hayes Record Shop received 180 records of the new top hit record. By the end of the day, 85% had been sold. How many of the records were sold?

9. Antonio is the chief photographer for a local magazine. He purchased a new camera for $350.40. His state has a 6% sales tax and a 1.5% city tax. How much tax did Antonio pay?

10. An answer to a problem computed on a calculator is 1.325. Write this answer as a percent.

ADVANCED APPLICATIONS

Calculate the answers to the problems given below.

1. U.S. experts agree that an important first step toward capitalism in former Communist controlled countries is the establishment of a private commercial banking system. As a result, First Bank opened a private bank in Poland so that private companies could borrow money. The bank is 80% U.S. owned. The remainder is owned by the National Bank.
   a. What percent is owned by Poland's bank?

   b. The bank loaned $12,000 to assist farmers in exporting hams to the United States. If 73% of the money loaned came from American investors, how much money did American investors loan?

2. The Lorenzo family earned $45,000 last year. They spent 18% for food, 26% for housing, 15% for clothing, 5% for entertainment and travel, 7.5% for car expenses, 6.5% for medical, and 16% for miscellaneous expenses.
   a. How much did they spend for each?

      Food

      Housing

      Clothing

      Entertainment and Travel

Car Expenses ______

Medical ______

Miscellaneous Expenses ______

b. What percent did they save? ______

c. What amount was saved? ______

3. A chef for The Sweet Shop ordered 40 kilograms of sugar to use in baking for the week. 5.7 kilograms were used on Monday, 6.6 on Tuesday, 5.2 on Wednesday, and 8.4 on Thursday.

a. If 78% of the sugar was used by the end of the day of Friday, how many kilograms were used on Friday? ______

b. How many pounds were used during the week? ______

unit 8 Percentages

"Nothing is worth doing wrong for."

Percents are found in many areas of business math. **Percent** is another way to state a decimal, fraction, or mixed number. You need to understand percents, how to convert them to another form, and how to use them in business situations.

After studying this unit, you should be able to:

1. use the percentage, base, and rate formulas to find a percent of a number, to find the percentage, and to find the base
2. find the percent increase or decrease
3. use the percentage formulas to solve business problems

MATH TERMS AND CONCEPTS

The following terminology is used in this unit. Become familiar with the meaning of each term so that you understand its usage as you develop math skills.

1. **Base**—the whole.
2. **Percent**—per (by) cent (hundredths); per hundred. Indicates a comparison of a number to 100.
3. **Percentage**—the part of the base (whole).
4. **Rate**—the percent used in relationships of numbers.

Identifying the Percentage, Rate, and Base

As you learned in Unit 7, percents establish relationships or comparisons between numbers. If you had a $10 bill and one half of it belonged to Brent, then $5 belongs to you and $5 belongs to Brent. Since one half is 50% when it is converted to a percent, 50% belongs to Brent and 50% belongs to you.

In this relationship of numbers, three elements are involved. First is the starting point of $10. This number represents 100% or the *whole* amount. This number (10) is called the **base**.

Second is the *percent* involved. It can be more or less than the base. In the above example, 50% is compared to $10. This number (50%) is called the **rate**.

Third is the *part* of the base when the rate is applied to the base. If the rate is less than 100%, the part will be less than the base. If the rate is greater than 100%, the part will be greater than the base. In the above example, $5 is the part. The part is called the **percentage**.

** Do not confuse percentage and percent. Percentage is the part taken from the base. Percent is the rate of that part and will always have the percent sign (%) following it.*

Letters are used in this book to represent the three terms:

Base (whole) = B
Rate (percent) = R
Percentage (part) = P

The key in solving any percentage problem is to first identify the three elements. In most percentage problems, the key words used to identify the three elements are *is* and *of*. The word *of* can precede the *base* in every situation. The word *is* will usually precede or follow the *percentage*. The *percent sign* (%) always means the rate.

Example: 8 is 25% of 32
P R B

In the following statements, identify each element by writing each number in the correct column.

	P	R	B
1. 159.3 is 27% of 590	______	______	______
2. 70% of 830 = 581	______	______	______
3. 1,256.25 is 37.5% of 3,350	______	______	______
4. 110 is 275% of 40	______	______	______
5. 5% of 100 is 5	______	______	______
6. 20.349 = 35.7% of 57	______	______	______
7. 19% of 4,000 = 760	______	______	______
8. 125% of 80 = 100	______	______	______
9. 56.772 = 68.4% of 83	______	______	______

Types of Percentage Problems

In problem-solving situations, *one* of the three elements will be missing. First identify the known elements. Then determine which element is missing.

Example: Find the *percentage* or part of a whole.

40% of 25 = ?
R B P

Find the *rate* or percent.

10 = ?% of 25
P R B

Find the *base* or whole (100%).

10 is 40% of ?
P R B

In the following problems, identify the three elements. Use a question mark for any missing term.

	P	R	B
1. 7% of 3.49 = ?	________	________	________
2. 7 is what percent of 19?	________	________	________
3. 20% of what is 140?	________	________	________
4. The Smith Company purchased 9,000 soccer balls. They sold 1,323 soccer balls. What percent were sold?	________	________	________
5. Department A pays $142 for custodial expenses, which is 5% of its monthly earnings. What are the monthly earnings for Department A?	________	________	________
6. The Harvee Company must pay 7% sales tax on the store's profit. If the store's profit is $27,210, how much is the sales tax?	________	________	________

Formulas

The following formulas are used to solve for the missing element in a percentage problem:

$$\text{Percentage} = \text{Base} \times \text{Rate}$$
$$\text{Base} = \text{Percentage} \div \text{Rate}$$
$$\text{Rate} = \text{Percentage} \div \text{Base}$$

This percentage triangle may help you to remember the percentage formulas:

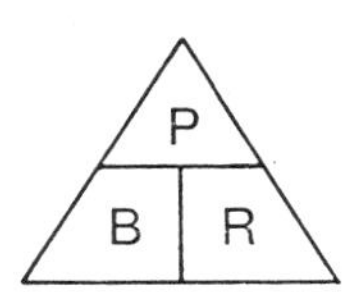

When one item is missing, you can look at the triangle and easily determine whether you should multiply or divide. When the percentage is missing, think of a multiplication problem:

$$P = B \times R \quad \textit{or} \quad P = R \times B$$

When the rate or the base is missing, think of a fraction: $\frac{P}{B}$ or $\frac{P}{R}$. Then divide the numerator by the denominator.

$$R = P \div B \quad \textit{or} \quad B = P \div R$$

Finding the Percentage or Part

When the percentage is missing, multiply the base times the rate.

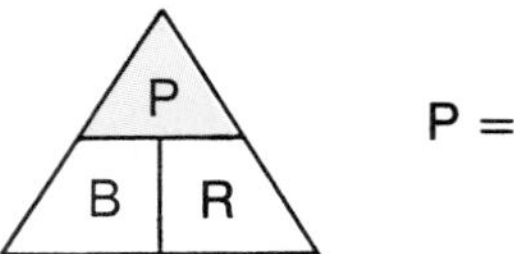

$$P = B \times R$$

A percent (rate) expresses a relationship between two numbers (the part and the whole); therefore, a percent of a number equals the part.

Example: 3% of 135 = a part of the whole (135)
= .03 × 135 = 4.05 (the part)

In this example, you are finding the part of the whole represented by 3%.

* ***If your calculator does not have a % key, change each percent to its decimal equivalent!***

Solve for the percentage in the following problems.

Decimal Selector on 2.

1. 5% of 60 = ____________ **2.** 25% of 95 = ____________

3. 109% of 34 = ____________ **4.** 87% of 100 = ____________

5. 35.7% of 72 = ____________ **6.** 120% of 230 = ____________

7. 40.25% of 1,400 = ________ **8.** 7% of 87 = ____________

9. 200% of 73 = ____________ **10.** 64% of 2,379 = ____________

The words *more than, greater than, less than,* or *smaller than* may be used near the base. To find the percentage, you multiply. If the problem reads *more than (greater than)* the base, the given rate is added to 100% before multiplying. If the problem reads *less than (smaller than)* the base, the given rate is subtracted from 100% before multiplying.

Examples: What amount is 5% *more than* 40?

100% + 5% = 105%
40 × 105% = 40 × 1.05 = 42

42 is 5% greater than 40.

What amount is 5% *less than* 40?

100% − 5% = 95%
40 × 95% = 40 × .95 = 38

38 is 5% less than 40.

Solve for the percentage in the following problems.

11. What amount is 9% more than 600? ____________

12. What amount is 9% less than 600? ____________

13. What amount is 30% more than 300? ____________

14. What amount is 30% less than 300? ____________

15. 400 increased by 7% is ? ____________

16. 400 decreased by 7% is ? ____________

Finding the Rate or Percent

When the rate is missing, a problem is stated in *one* of three ways:

1. 11.35 *is* what percent *of* 227?
2. Find what percent 11.35 *is of* 227.
3. What percent *of* 227 *is* 11.35?

To find the rate or percent, divide the percentage by the base.

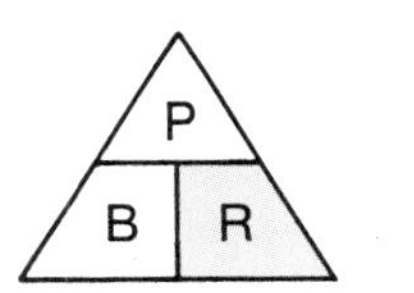

R = P ÷ B

Use the amount following the word *of* as the divisor and change the decimal answer to a percent.

Example: 11.35 is what percent of 227?

11.35 ÷ 227 = .05 = 5%

In each of the following problems, find the percent. Remember to give the answer in percent form. * ***Round each percent to two decimal places.***

Use the % key instead of the equals key to convert your answers to percents.

1. What percent of 40 is 8? ____________

2. 77 is what percent of 122? ____________

3. Find what percent 12 is of 35. ____________

4. What percent of 67 is 17?

____

5. 6 is what percent of 24?

____

6. Find what percent 27 is of 38.

____

Again, the words *more than, greater than, less than,* or *smaller than* may be used near the base. To find the rate, you divide. However, the percentage that matches the rate you are finding is the *difference* between the percentage given and the base. Therefore, subtract the smaller number from the larger number and then divide by the base.

Examples: 3 is what percent *less than* 15?

15 − 3 = 12
12 ÷ 15 = 80% (the 12 and 80% match)

20 is what percent *more than* 15?

20 − 15 = 5
5 ÷ 15 = 33.33% (the 5 and 33.33% match)

Solve for the rate in these problems.

7. 8 is what % greater than 6?

____

8. 14 is what % less than 20?

____

9. What % less than 60 is 45?

____

10. What % more than 40 is 55?

____

Are you using aliquot parts?

11. 60 is what % more than 20?

____

12. 6 is what % less than 24?

____

Finding the Base or Whole

When the base is missing, the problem is stated in *one* of three ways:

1. 5% of what amount *is* 11.35?
2. 11.35 *is* 5% of what amount?
3. Find an amount of which 5% *is* 11.35.

To find the base, divide the percentage by the rate.

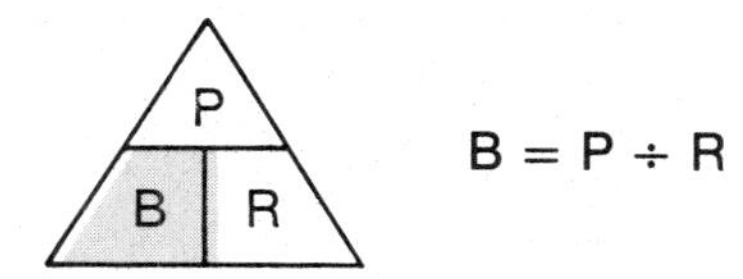

Use the percent key in place of the equals key. The operation will convert the rate to a decimal and divide.

Example: 11.35 is 5% of what amount?

11.35 ÷ .05 = 227

Solve for the base in these problems. Round to two places.

1. 127 is 19% of ? ____________

2. 7% of what is 32? ____________

3. 140 is 44% of ? ____________

4. $12\frac{1}{2}$% of what is 242? ____________

5. 22% of what is 130? ____________

6. 90% of ? is 310? ____________

Again, the words *more than, greater than, less than,* or *smaller than* may be used in a percentage problem. To find the base, divide the percentage by the rate. If the problem reads *more than (greater than)* the base, the given rate is added to 100% before you divide. If the problem reads *less than (smaller than)* the base, the given rate is subtracted from 100% before dividing.

Examples: 80 is 20% *more than* what number?

100% + 20% = 120%
80 ÷ 1.2 = 66.67

80 is 20% *less than* what number?

100% − 20% = 80%
80 ÷ .8 = 100

Solve for the base in the following problems. * *Round to two places.*

7. 680 is 35% more than what number? ____________

8. 490.5 is 27% less than what number? ____________

9. 36 is 25% more than what number? __________

10. 10 is 5% less than what number? __________

11. 30 is 15% more than what number? __________

12. 32 is 14% less than what number? __________

Finding the Percent Increase or Decrease

When comparing one amount with another, businesses look at the *percent increase* or *decrease* in amounts. For example, a business might compare one year's utility expenses with the previous year's utility expenses.

To determine the percent increase or decrease, find the amount of the increase or decrease by subtracting the smaller amount from the larger amount. This amount is the percentage. The original amount, or earliest amount, becomes the base. You are solving for the rate or percent.

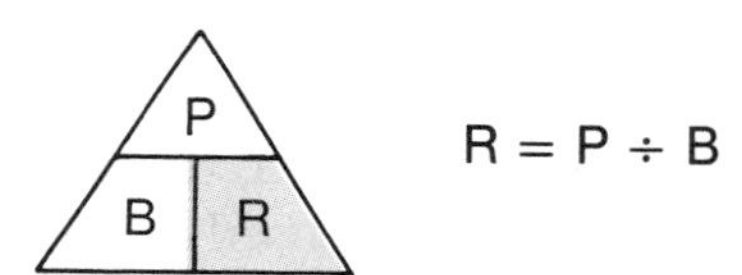

Examples: Assume a business wants to compare utilities from 1992 to 1993. Utilities in 1992 were $7,257 and utilities in 1993 were $8,640. What is the percent increase?

$8,640 – $7,257 = $1,383 (amount of increase)
$1,383 ÷ $7,257 = .1906 = 19.06%

* ***Remember, the original (earliest) amount is the base.***

Assume a business wants to compare utilities from 1990 to 1992. Utilities in 1992 were $7,257 and utilities in 1990 were $8,781. What is the percent decrease?

$8,781 – $7,257 = $1,524 (amount of decrease)
$1,524 ÷ $8,781 = .1736 = 17.36%

For each of the following problems find the percent increase or decrease. Use parentheses to indicate the amount of decrease and the percent decrease.

* ***Round each percent to two decimal places. Some calculators have a delta key. The delta key will compute the percent increase or decrease. Refer to Appendix A.***

1. Personal Income by Regions (Per Individual)

	1986	1987	Amount Increase/Decrease	Percent Increase/Decrease
Northeast	$14,526	$14,934	______	______
Midwest	$12,514	$12,626	______	______
South	$11,486	$11,599	______	______
West	$13,507	$13,636	______	______

2. Expenses by Department

Department	1991	1992	Amount Increase/Decrease	Percent Increase/Decrease
Sales	$235,567	$204,789	______	______
Advertising	$78,459	$81,356	______	______
Office	$196,581	$167,234	______	______

3. Amount of revenue projected for Cable TV Industry for 1990 and 1995 (Stated in millions of dollars)

1990	1995	Amount Increase/Decrease	Percent Increase/Decrease
$15,299	$23,434	______	______

Average monthly basic rate:

1990	1995		
$16.64	$24.88	______	______

Solving Word Problems

Before solving a word problem, be sure the numbers given in the problem match the desired solution.

Example: A store purchased 60 lawn mowers. Ten percent of the lawn mowers had defective parts and were returned. The rest of the lawn mowers were sold. How many lawn mowers were sold?

We can look at the example and see that 60 must be the base or 100% of the lawn mowers purchased. Ten percent of the mowers were returned. If the example asked for the *number* of mowers returned, the formula $P = B \times R$ could be used to get the percentage. But the percent given and the element we want to find do not match. To solve the problem, you must determine what percent of the mowers were sold by subtracting the given rate from 100%. Then solve for the percentage using the new rate.

Example: $100\% - 10\% = 90\%$ (percent of mowers sold)

$B \times R = P$

$60 \times .9 = 54$ (number of mowers sold)

By permission of Johnny Hart and NAS, Inc.

Finding the missing element may not be the answer to the problem. You may have to add or subtract the missing element to another number in the problem to get the answer.

Example: Josefina jogged two miles. She has jogged 25% of the total distance of her goal. How much farther must Josefina jog? In this example, you must first find the total distance (the base) that Josefina is going to jog. Then subtract the two miles she has already jogged from the base to find how much farther she must jog.

$P \div R = B$

$2 \div .25 = 8$ (the base)

$8 - 2 = 6$ (how much farther Josefina needs to jog)

* ***Remember, the word "of" will precede the base. The rate is identified by the percent sign. Be sure the percentage and the rate match before trying to solve any word problem.***

The percentage may be more or less than the base. If the percentage is less than the base, then the rate will be less than 100%. If the percentage is more than the base, the rate will be more than 100%.

* ***Remember, you work problems that are stated as more than or less than the base just as your work percent increase or decrease problems.***

EVALUATING YOUR SKILLS

The following problems evaluate your skills in the material covered in Units 1 through 7. Some of the problems will evaluate your ability to work problems mentally while others will evaluate your ability to use your calculator. Work the problems as quickly as you can.

1. Round 654,621.78 to the nearest thousand. ____________

2. Round $56.4537 to the nearest hundredth. ____________

3. Write 1.762 as a percent. ____________

4. Write .43 as a fraction. ____________

5. $4{,}261 \div 100 = ?$ ____________

6. Write $\frac{3}{4}$ as a percent. ____________

7. Estimate the quotient of $662{,}197.88 \div 55$. ____________

8. $\frac{3}{8}$ is what aliquot part? ____________

9. $3.05 \times .527 = ?$ ____________
* *Round to three decimal places.*

10. $11.24 + 463.78 - 221.00 + 59.80 - 148.66 = ?$ ____________

11. $561.37 - 802.40 = ?$ ____________

12. Use the cancellation method to multiply: $\frac{4}{9} \times \frac{27}{36}$ ____________

13. Find the least common denominator for $\frac{5}{9} + \frac{5}{6} + \frac{5}{24}$ ____________

14. $(574.32 \times .67) + 23 = ?$ ____________

15. A ticket seller started with 240 seventy-five cent tickets, 75 fifty cent tickets, and $25 in cash to make change. At the end of the game, the ticket seller returned 20 seventy-five cent tickets, 18 fifty cent tickets, and $241.50 in cash.

a. Was the ticket seller short or over? ____________

b. By how much? ____________

APPLYING YOUR KNOWLEDGE

Using the information given, calculate the correct answer. Determine what elements are given and what you must find. * *Round answers to two decimal places.*

1. Property valued at $72,000 was sold. The commission was $4,320. What was the percent commission?
   a. What is the base? ______
   b. What is the percentage? ______
   c. What element is missing? ______
   d. What is the percent commission? ______

2. A salesperson earns commissions of 33% of gross profit on total sales. If total sales for the week amounted to $7,800, how much did she earn?
   a. What is the base? ______
   b. What is the rate? ______
   c. What element is missing? ______
   d. What is the amount earned? ______

3. Five departments pay utility expenses at the listed percents of total monthly earnings. What is the total monthly earnings for each department?

Department	Utilities	Percent	Total Monthly Earnings
A	$235	7%	______
B	$93	3%	______
C	$324	19%	______
D	$215	16%	______
E	$212	14%	______

4. An investor subscribed to Thin Corporation's stock. He paid $425,000 down. This was 75% of the purchase price of the stock. What is the purchase price?
   a. What is the rate? ______
   b. What is the percentage? ______
   c. What element is missing? ______
   d. What is the purchase price? ______

5. The price of gasoline decreased from $4.84 per liter to $3.96 per liter. What is the percent decrease?
   a. What is the base?
   b. What is the percentage?
   c. What element is missing?
   d. What is the percent decrease?

6. The average cost of a house decreased from $78,000 to $56,000. What is the percent decrease?
   a. What is the base?
   b. What is the percentage?
   c. What element is missing?
   d. What is the percent increase?

7. The Toggle and Bolts Shop rents space in a shopping mall. They must pay 11% of their gross profit per year for rent. If their gross profit is $132,000, how much rent do they owe?
   a. What is the base?
   b. What is the rate?
   c. What element is missing?
   d. How much rent is due?

8. Sales tax revenue for the city decreased from $3,467,298 in 1990 to $2,987,456 in 1991. What is the percent decrease?
   a. What is the base?
   b. What is the percentage?
   c. What element is missing?
   d. What is the percent decrease?

9. The Johnson's have determined that they will need to receive $8,945 per year from investments in addition to income from pensions to cover their living expenses during retirement. If they can receive 9% interest from an investment, how much would they need to invest?
   a. What is the rate?
   b. What is the percentage?
   c. What element is missing?
   d. How much do they to invest?

ADVANCED APPLICATIONS

Using the information given, solve the problems.

1. How much will a savings deposit of $80 be worth at the end of one year if it earns $6\frac{3}{4}\%$ interest?

   ____

2. What is Gene's earnings if he earns 40% more than Alice and Alice earns $378?

   ____

3. What was the original price of a bicycle if it sold for $135 after a 25% discount?

   ____

4. How much must you sell to earn $1,750 if you receive 6% of sales?

   ____

5. Seminole Cleaner's rent for April is $850. This is 12.5% more than last year's rent. What was last year's rent? (Round to nearest whole dollar.)

   ____

6. The Wholesale Shop pays $10,200 annually for rent. Beginning next month their rent will increase to $950 per month. What is the percent increase?

   ____

7. Stephen's Camera Shop is opening an additional store. The business earned $678,000 in sales this year. They expect next year's sales to increase by 28%. What will be projected sales for next year?

   ____

The Hodgepodge Shop

series II A Retail Application

"Do not give anyone a piece of your mind until you are very sure you can spare it."

Welcome back to The Hodgepodge Shop. You will continue in the training program by working with inventory control maintaining inventory and purchasing records.

After completing the inventory and purchasing training program, you should be able to:

1. verify, extend, and record information on a merchandise requisition form
2. prepare merchandise authorization forms
3. extend purchase orders
4. calculate the percent increase/decrease of sales and inventory

TERMINOLOGY

The following terminology is used in this application. Become familiar with the meaning of each term so that you understand its usage.

1. **Data Entry**—the recording of information to feed into a computer.
2. **Inventory**—merchandise on hand.
3. **Merchandise Authorization (MA)**—a form used by businesses to show approval for purchasing merchandise.
4. **Merchandise Requisition (MR)**—a form used by businesses to list merchandise needed and to record the merchandise shipped to the store.

In this segment of your training, you will be helping with the reordering process for the four stores (the downtown store and the three branch stores). The reordering process is a part of inventory control. Computers have made the process of inventory control a more accurate and faster procedure. Read each job assignment carefully. Follow each step outlined for you. Do not skip any steps. Accurately prepare the forms to send to the **data entry** department for inventory update.

Good luck! Remember, you are a trainee, so ask your supervisor for help as you need it.

Assignment 1: MA and MR

When one of The Hodgepodge Shop's stores needs to add inventory, the store manager completes four copies of a **Merchandise Requisition (MR)**. The manager includes on the MR the catalog number, item description, packing units, quantity, and packing unit price for each item. The MR is then sent to the main office.

The Hodgepodge Shop operates within a budget. The main office checks to see that the cost of the merchandise requested falls within the amount budgeted before authorizing the warehouse to fill any orders. The warehouse cannot ship the merchandise to the store unless a **Merchandise Authorization (MA)** is attached to the MR.

At the main office, you will verify catalog numbers and prices on the MR before sending a copy to the warehouse. The main office has received merchandise requisitions from Stores #2 and #3.

January 3

1. Remove the Catalog (Form 1, page 157) and the Merchandise Requisitions from Stores #2 and #3 (Forms 2 and 3). Using the catalog, verify each item number and packing unit price on the MR forms. If an error was made, draw a line through the figure and write the correct figure above.
2. Extend the unit price and quantity ordered. Write the extension in Column 6 of the MR. * ***Compute the price per unit. Then compute the price for the number ordered. Set the decimal selector on 4 and mentally round total price to two places.***
3. Total and verify Column 6.
4. Remove the Merchandise Authorizations (Forms 4 and 5). Complete the two Merchandise Authorization Forms for Store #2 and Store #3 using the following information:

   Write in the store number. Use January 3 as the date. Sign your name after *Authorized by.* This will be initialed by your supervisor. Write "To replenish stock" after reason. Write the total of Column 6 from the store's MR after *Cost of Merchandise Requested.* File the MA's for future use. (A copy of the MR and MA is sent to the warehouse.)

Assignment 2: Packing Slips

Once the warehouse receives an MR, a packing slip is prepared in triplicate for the merchandise in stock. One copy of the packing slip is sent to the store along with the merchandise. One copy is sent to the main office, and one copy is kept for the warehouse files.

When the main office receives the packing slips from the warehouse, the merchandise sent to the stores is recorded on the MR.

You have received copies of packing slips for merchandise sent directly to the stores from the warehouse. Some of the merchandise was not in stock and had to be ordered from the wholesaler. These items are labeled "backordered" on the packing slips.

January 4

1. Remove the Packing Slips (Forms 6 and 7) and separate them by store number. On the Merchandise Requisitions (Forms 2 and 3), record the quantity of each item shipped in Column 7 and the quantity backordered in Column 9. File the packing slips.
2. Extend the quantity shipped in Column 7 using the packing unit price in Column 5. Write the extension in Column 8.
3. Total and verify Column 8.
4. Remove the Merchandise Authorizations (Forms 4 and 5). Write the total of Column 8 beside *Cost of Merchandise Shipped* for each store. File for future use. (A copy of the MA and MR forms would be sent to data entry to update inventory.)
5. File the MR forms for later use.

Assignment 3: Purchase Orders

The warehouse prepared purchase orders for merchandise not in stock. The supplier's name and address, the item, the quantity, and the description were typed on the purchase orders.

The main office received copies of these purchase orders prepared by the warehouse. The quantity ordered is the merchandise ordered for the stores as well as for restocking the warehouse. You will extend and total purchase orders and notify data entry of changes in inventory.

January 5

Use grand total to total all purchase orders.

1. Remove the Purchase Orders (Forms 8 through 12). Fill in the prices using the catalog (Form 1). File the catalog.
2. Extend the quantity and price column to the total column. Then total each purchase order.
3. Verify the sums of all purchase orders.
4. Sign the purchase orders.
5. Remove the Purchase Requisition (Form 13). Write the grand total for the purchase orders on the Purchase Requisition Form. Write in the date, and sign your name. Your supervisor should initial the form.
6. Submit the Purchase Requisition and Purchase Orders to your supervisor who will mail the original and send the copies to data entry.

Assignment 4: Packing Slips

The warehouse received the merchandise ordered on January 5. The warehouse examined the merchandise, prepared packing slips, and sent the merchandise directly to the stores.

A notice was included in the merchandise received for a price increase on some items. This increase has been noted on the packing slips.

Some of the merchandise is no longer available and is indicated by "discontinued" on the packing slips. You received a copy of the packing slips at the main office.

January 19

1. Remove the Packing Slips (Forms 14 and 15) and the Merchandise Requisitions (Forms 2 and 3). From the packing slips, record the quantity shipped in Column 10 on the MR for Stores #2 and #3.
2. If there is a price increase, compute the increase on unit price. Round up and record the new price in Column 11. Extend the price from Column 5 to Column 11 if there is no change. File the packing slips.
3. Extend Columns 10 and 11 to Column 12 on Forms 2 and 3.
   * ***Remember to compare the packing units in Column 3 with the quantity shipped in Column 10 to compute the total price.***
4. Total and verify Column 12.
5. Remove the Merchandise Authorizations (Forms 4 and 5). Write the total of Column 12 on the MA's after *Backordered Merchandise Received and Shipped.*
6. Add the MA's total Cost of Merchandise Shipped and Backordered Merchandise Received and Shipped. Write the total after *Total Cost Charged to Store*. Submit the MA's and MR's to your supervisor. (A copy is sent to data entry.)

Assignment 5: Schedule of Comparative Sales

The Hodgepodge Shop carefully watches the change in sales from one month to another. You have been asked to calculate the percent increase/decrease for the four stores from December to January.

February 1

1. Remove Schedule of Comparative Sales (Form 16). Complete Form 16 and submit it to your supervisor.

   * ***Indicate decreases in sales and percents by parentheses.***

Assignment 6: Budget

Watching the budget is an important function at The Hodgepodge Shop. You have been asked to complete a budget analysis for Store #3 for October through December. This information will be used to prepare future budgets.

February 2

1. Remove Budget (Form 17). Find the percent of the total budget spent for each type of merchandise purchased and record on Form 17. Submit the budget to your supervisor.

* ***Remember the P-B-R formulas?***

SELF-EVALUATION

Were you satisfied with the self-evaluation you completed in Series I? Have you been trying to improve in your weak areas?

Your teacher may have you complete the self-evaluation again. If you still have areas in which you would like to improve, work hard to do so. In a later training session you will be given an employer's job evaluation. The evaluation is used to determine who is offered full-time employment with The Hodgepodge Shop. It is important that you work to improve yourself if you want to move up the ladder of success.

CATALOG

Form 1

Item No.	Item Description	Standard Package	Packing Unit Price	Suggested List Price
Southern Stationery Supply Co., 700 9th Ave., New York, NY 10019-4322				
Photo Albums				
sx-B1	Burgundy	3	30.00	42.00
sx-B2	Oyster White	3	30.00	42.00
sx-B4	Blue Heaven	3	30.00	42.00
sy-T9	Teddy Bears	3	37.50	54.00
sy-T8	All Sports	2	23.60	34.00
sy-T0	Daisies	2	23.60	34.00
sw-C1	Bonded Leather (Brown)	1	19.00	27.50
sw-C2	Bonded Leather (Black)	1	19.00	27.50
Snapshot Books				
CR-88	Grandma's Snapshots	3	13.50	19.50
CR-45	Memories	3	13.50	19.50
CR-54	Our Baby	3	13.50	19.50
NP-04	Burgundy	4	20.00	30.00
NP-17	Brown	4	20.00	30.00
Scrapbooks				
Pn14	Ducks	5	62.50	90.00
Pn17	Ballet Dancers	5	62.50	90.00
Pn18	Football	5	62.50	90.00
P18	Brown	5	52.00	75.00
P15	Red	5	52.00	75.00
S.R. Paper Products, 3024 Excellsior Blvd., Minneapolis, MN 55416-7091				
Rock 'n Roll Pattern				
CR1-11	12″ Dinner Plates (pkg of 10)	doz	35.00	51.00
CR2-11	Luncheon Plates (pkg of 10)	12	30.00	42.00
CR2-12	Dinner Napkins (pkg of 24)	10	18.50	24.00
CR2-14	Luncheon Napkins (pkg of 24)	$\frac{1}{2}$ doz	11.40	16.50
CR2-15	Beverage Napkins (pkg of 20)	$\frac{1}{2}$ doz	16.00	20.00
CR2-17	9 oz cups (pkg of 12)	10	15.50	22.50
CR2-18	Invitations (pkg of 8)	8	18.00	26.00
CH-22	Tablecloth (one per pkg)	5	3.00	4.32

Item No.	Item Description	Standard Package	Packing Unit Price	Suggested List Price
Runsey's Supply House, 39 N. 8th St., Minneapolis, MN 55416-7091				
Rose Garden Desk Accessories Selection				
BX11-9	Desk Pads	3	41.00	58.50
BX11-4	Pencil Cups	3	12.60	18.00
BX12-0	Letter Holders	3	23.00	33.00
BX12-6	Letter Openers (Brass)	3	28.00	40.50
BX66-2	Memo Paper with Holder	1	7.00	10.00
BX72-5	Pen and Stand Sets	3	25.00	36.00
BX14-4	Memo Box	1	18.75	27.00
BX14-4A	Memo Box Refill	3	25.00	36.00
BX27-9	Letter Tray	1	39.99	56.00
BX8-66	Photo Frame	3	25.00	36.00
Goodman's Gift House, P.O. Box 22211, San Francisco, CA 94100-0385				
4321-2	Quartz Bookends	1	32.00	49.95
5210-2	Quartz Desk Clock	1	35.00	54.95
8742-4	Quartz Paperweight	2	11.35	17.90
Frames, Inc., 4102 Wayzata Blvd., Minneapolis, MN 55416-1031				
33471rt	Portrait Miniature Picture Frame, 4″ high	2	9.52	15.00
6792pq	Silverplated Picture Frame, 5×7	2	44.40	69.90
6884x1	Oval Brass Picture Frame, $3\frac{1}{2} \times 3$	3	22.75	35.85
6533xy	Crystal Picture Frame, $7\frac{1}{2} \times 5\frac{3}{4}$	2	35.50	55.90

Form 2

DATE 1/3/--	*The Hodgepodge Shop*		STORE 2			MERCHANDISE REQUISITION FORM					
			Merchandise Requested			Merchandise Shipped			Shipped		
Item No.	Item Description	Packing Units	Quantity	Packing Unit Price	Total Price	Quantity	Total Price	Back Ordered	Quantity	Packing Unit Price	Total
1	2	3	4	5	6	7	8	9	10	11	12
sw-C1	Bonded Leather/Brown-Ph.Al.	1	5	$ 19.00							
sw-C2	Bonded Leather/Black-Ph.Al.	1	6	19.00							
CR1-11	12″ Dinner Plates/R 'n Roll	1 doz	42	35.00							
CR2-15	Beverage Napkins/R 'n Roll	$\frac{1}{2}$ doz	36	16.00							
CR2-17	9 oz Cups/R 'n Roll	10	7	15.50							
CH-22	Tablecloth/R 'n Roll	5	22	3.00							
BX11-9	Desk Pads/Rose Garden Sel.	3	2	41.00	**		**				
BX12-0	Letter Holders/Rose Garden	3	21	23.00							
BX8-66	Photo Frame/Rose Garden	3	9	25.00							
BX27-9	Letter Tray/Rose Garden	1	11	39.99							
5210-2	Quartz Desk Clock	1	4	35.00							
8742-4	Quartz Paperweight	2	8	11.35						(Discontinued)	
	Column Totals				*		*				*

*DID YOU USE GT or MEMORY?
**Round up

DATE 1/3/--	*The Hodgepodge Shop*		STORE 3			MERCHANDISE REQUISITION FORM					
			Merchandise Requested			Merchandise Shipped			Shipped		
Item No.	Item Description	Packing Units	Quantity	Packing Unit Price	Total Price	Quantity	Total Price	Back Ordered	Quantity	Packing Unit Price	Total
1	2	3	4	5	6	7	8	9	10	11	12
sy-T8	All Sports/Photo Album	2	7	$ 23.60							
CR-45	Memories/Snapshot Book	3	7	13.50							
Pn18	Football/Scrapbook	5	7	62.50							
CR2-11	Luncheon Plates/R 'n Roll	12	2 doz	30.00							
CR2-14	Luncheon Napkins/R 'n Roll	$\frac{1}{2}$ doz	24	11.40							
CR2-18	Invitations/R 'n Roll	8	24	18.00							
BX11-4	Pencil Cups/Rose Garden	3	2	12.60							
BX12-6	Letter Openers/Brass	3	12	28.00							
BX14-4	Memo Box	1	9	18.75							
BX14-4A	Memo Box Refill	3	6	25.00							
33471rt	Portrait Picture Fr./4″	2	8	9.52						**	
6884x1	Oval Brass Pic. Fr./$3\frac{1}{2} \times 3$	3	15	22.75							
	Column Totals				*		*				*

*DID YOU USE GT or MEMORY?
**Round up

The Hodgepodge Shop

Form 4

MERCHANDISE AUTHORIZATION

Store ______________

Date ______________

Authorized By ____________________________ Supervisor's Initials ____________

Reason ____________________________

Cost of Merchandise Requested ____________________

Cost of Merchandise Shipped ____________________

Backordered Merchandise Received and Shipped ____________________

Total Cost Charged to Store ____________________

The Hodgepodge Shop

Form 5

MERCHANDISE AUTHORIZATION

Store ______________

Date ______________

Authorized By ____________________________ Supervisor's Initials ____________

Reason ____________________________

Cost of Merchandise Requested ____________________

Cost of Merchandise Shipped ____________________

Backordered Merchandise Received and Shipped ____________________

Total Cost Charged to Store ____________________

PACKING SLIP

Form 6

The Hodgepodge Shop
705 4th Ave. S.
Minneapolis, MN 55415-2232

DELIVER TO: The Hodgepodge Shop

LOCATION: St. Paul, MN 55191-8413

NO.: 910

STORE NO.: 2

DATE: January 4, 19--

QUANTITY	DESCRIPTION
5	sw-C1 Bonded Leather/Brown-Ph. Album
4	sw-C2 Bonded Leather/Black-Ph. Album (2-Backordered)
18	CR1-11 12" Dinner Plates/R 'n Roll (2 doz-Backordered)
1/2 doz	CR2-15 Beverage Napkins/R 'n Roll (2 1/2 doz-Backordered)
7	CR2-17 9 oz Cups/R 'n Roll
22	CH-22 Tablecloth
2	BX11-9 Desk Pads/Rose Garden Selection
12	BX12-0 Letter Holders/Rose Garden (9-Backordered)
9	BX8-66 Photo Frame/Rose Garden Selection
	BX27-9 Letter Tray/Rose Garden (11-Backordered)
4	5210-2 Quartz Desk Clock
	8742-4 Quartz Paperweight (8-Backordered)

PACKING SLIP

Form 7

The Hodgepodge Shop
705 4th Ave. S.
Minneapolis, MN 55415-2232

DELIVER TO: The Hodgepodge Shop

LOCATION: Minneapolis, MN 55422-5871

NO.: 911

STORE NO.: 3

DATE: January 4, 19--

QUANTITY	DESCRIPTION
7	sy-T8 All Sports/Photo Album
7	CR-45 Memories/Snapshot Book
3	Pn18 Football/Scrapbook (4-Backordered)
2 doz	CR2-11 Luncheon Plates/R 'n Roll
1/2 doz	CR2-14 Luncheon Napkins/R 'n Roll (18-Backordered)
24	CR2-18 Invitations/R 'n Roll
2	BX11-4 Pencil Cups/Rose Garden Selection
3	BX12-6 Letter Openers/Brass (9-Backordered)
9	BX14-4 Memo Box
6	BX14-4A Memo Box Refill
4	33471rt Portrait Picture Frame/4" (4-Backordered)
15	6884x1 Oval Brass Picture Frame/3 1/2 x 3

PURCHASE ORDER

Form 8

The Hodgepodge Shop
705 4th Ave. S.
Minneapolis, MN 55415-2232

Southern Stationery Supply Co.
700 9th Ave.
New York, NY 10019-4322

PURCHASE ORDER NO.: 1319

DATE: January 5, 19--

TERMS: 2/10, n/30

SHIP VIA: Truck

QUANTITY	CAT. NO.	DESCRIPTION	PRICE	TOTAL
2	sw-C2	Bonded Leather/Black - Photo Album		
1 pkg	Pn18	Football/Scrapbook		
		Total		

BY ______________________________ PURCHASING AGENT

PURCHASE ORDER

Form 9

The Hodgepodge Shop
705 4th Ave. S.
Minneapolis, MN 55415-2232

S.R. Paper Products
3024 Excellsior Blvd.
Minneapolis, MN 55416-7091

PURCHASE ORDER NO.: 1320

DATE: January 5, 19--

TERMS: 2/10, n/30

SHIP VIA: Truck

QUANTITY	CAT. NO.	DESCRIPTION	PRICE	TOTAL
2 doz	CR1-11	12" Dinner Plates/R 'n Roll		
10 pkg	CR2-15	Beverage Napkins/R 'n Roll		
3 pkg	CR2-14	Luncheon Napkins/R 'n Roll		
		Total		

BY ______________________________ PURCHASING AGENT

PURCHASE ORDER Form 10

The Hodgepodge Shop
705 4th Ave. S.
Minneapolis, MN 55415-2232

Runsey's Supply House
39 N. 8th St.
Minneapolis, MN 55416-7091

PURCHASE ORDER NO.: 1321

DATE: January 5, 19--

TERMS: 2/10, n/30

SHIP VIA: Truck

QUANTITY	CAT. NO.	DESCRIPTION	PRICE	TOTAL
3 pkg	BX12-0	Letter Holders/Rose Garden		
11	BX27-9	Letter Tray/Rose Garden		
3 pkg	BX12-6	Letter Openers/Brass		
		Total		

BY ______________________ PURCHASING AGENT

PURCHASE ORDER Form 11

The Hodgepodge Shop
705 4th Ave. S.
Minneapolis, MN 55415-2232

Goodman's Gift House
P.O. Box 22211
San Francisco, CA 94100-0385

PURCHASE ORDER NO.: 1322

DATE: January 5, 19--

TERMS: 2/10, n/30

SHIP VIA: Truck

QUANTITY	CAT. NO.	DESCRIPTION	PRICE	TOTAL
4 pkg	8742-4	Quartz Paperweight		
		Total		

BY ______________________ PURCHASING AGENT

PURCHASE ORDER

Form 12

The Hodgepodge Shop
705 4th Ave. S.
Minneapolis, MN 55415-2232

Frames, Inc.
4102 Wayzata Blvd.
Minneapolis, MN 55416-1031

PURCHASE ORDER NO.: 1323

DATE: January 5, 19--

TERMS: 2/10, n/30

SHIP VIA: Truck

QUANTITY	CAT. NO.	DESCRIPTION	PRICE	TOTAL
2 pkg	33471rt	Portrait Picture Frame/4"		
		Total		

BY ______________________ PURCHASING AGENT

The Hodgepodge Shop

PURCHASE REQUISITION

Form 13

Date ______________________

Total Amount Purchased ______________________

Authorized By ______________________

Supervisor's Initials ______________________

PACKING SLIP

Form 14

The Hodgepodge Shop
705 4th Ave. S.
Minneapolis, MN 55415-2232

DELIVER TO: The Hodgepodge Shop

LOCATION: St. Paul, MN 55191-8413

NO.: 912

STORE NO.: 2

DATE: January 19, 19--

QUANTITY	DESCRIPTION
2	sw-C2 Bonded Leather/Black - Photo Album
2 doz	CR1-11 12" Dinner Plates/R 'n Roll
2 1/2 doz	CR2-15 Beverage Napkins/R 'n Roll
9	BX12-0 Letter Holders/Rose Garden Selection
11	BX27-9 Letter Tray/Rose Garden Selection
(discontinued)	8742-4 Quartz Paperweight
	Price Increase - Effective January 15
	7% increase on all paper products
	10% increase on all frames

PACKING SLIP

Form 15

The Hodgepodge Shop
705 4th Ave. S.
Minneapolis, MN 55415-2232

DELIVER TO: The Hodgepodge Shop

LOCATION: Minneapolis, MN 55422-5871

NO.: 913

STORE NO.: 3

DATE: January 19, 19--

QUANTITY	DESCRIPTION
4	Pn18 Football/Scrapbook
18	CR2-14 Luncheon Napkins/R 'n Roll
9	BX12-6 Letter Openers/Brass
4	33471rt Portrait Picture Frame/4"
	Price Increase - Effective January 15
	7% increase on all paper products
	10% increase on all frames

The Hodgepodge Shop

Form 16

SCHEDULE OF COMPARATIVE SALES

Store	December Monthly Sales	January Monthly Sales	Increase/Decrease	Percent Increase/Decrease
#1	$52,372	$42,364	________	________
#2	$60,286	$70,500	________	________
#3	$62,481	$65,692	________	________
#4	$71,364	$78,255	________	________

* *Round to two decimal places.*

The Hodgepodge Shop

Form 17

BUDGET ANALYSIS FOR OCTOBER-DECEMBER

Store #3

Total Amount Budgeted $93,888

Merchandise	Amount Spent	Percent of Total Budget
XL-1 Paper Products	$24,701.93	________
ST-2 Stationery Items	$27,819.01	________
CD-3 Cards, Gift Wrap	$4,422.12	________
AC-4 Accessories	$2,995.03	________
GF-5 Gifts	$15,069.02	________
Total	________	________

Did you use constant division?

What percent of the total amount budgeted was not spent?

____

unit 9 Discounts

"Of all the things you wear, your expression is the most important."

Buying merchandise to sell to the consumer is a major function of business. The successful retailer makes purchases wisely, taking advantage of opportunities to save money. This means comparing prices among various manufacturers, getting the best trade discounts, and taking available cash discounts.

After studying this unit, you should be able to:

1. calculate cash discounts and invoice price
2. determine the due date
3. compute single and series trade discounts
4. calculate and use complements in single and series discounts
5. calculate single discount equivalents

MATH TERMS AND CONCEPTS

The following terminology is used in this unit. Become familiar with the meaning of each term so that you understand its usage as you develop math skills.

Terms are defined based on how they are used in this unit.

1. **Cash Discount**—an amount subtracted from the sales price if the invoice is paid by a certain date.
2. **Complement**—the difference between 100% and the discount rate.
3. **Discount**—an amount that reduces the full price of an item.
4. **Discount Rate**—a percent.
5. **Invoice**—a bill showing the amount due for the merchandise along with shipping and billing information.
6. **Invoice Price**—amount to be paid on an invoice; net price less deductions.
7. **List Price**—the manufacturer's suggested retail price on products sold to retail customers.
8. **Net Price**—list price less discount.
9. **Prepaid Freight**—freight charges that are paid by the seller before merchandise is shipped.
10. **Price List**—a list of prices of items for sale to retailers and wholesalers.
11. **Series Discounts**—more than one discount.
12. **Terms**—the conditions under which a discount is offered.
13. **Trade Discount**—a discount deducted from the list price and given to wholesalers and retailers.

The merchandise ordered by a retailer is recorded on an **invoice**. To encourage prompt or early payment of an invoice, a wholesaler may offer **cash discounts**.

Terms

The **terms** of a cash discount state when an invoice must be paid if a **discount** is to be taken. This time period is a certain number of days from the date of the invoice. If an invoice is not paid within the discount time period, the terms will state when the full amount of the invoice is due. Terms can be expressed in several ways.

1. *2/10,n/30*—Read: two-ten-net-thirty.
   A 2% discount is given if an invoice is paid within 10 days of the date on the invoice. If it is not paid within 10 days, the *net amount* is due within 30 days.
2. *3/10,2/15,n/30*—Read: three-ten-two-fifteen-net-thirty.
   A 3% discount is given if an invoice is paid within 10 days of the date of an invoice. A 2% discount is given if an invoice is paid within 15 days of the date of an invoice. The net amount is due within 30 days.
3. *2/10 EOM*—Read: two-ten-end of month.
   A 2% discount is given if an invoice is paid within 10 days after the end of the current month. The net amount is due at the end of the next month.
4. *n/30*—Read: net-thirty.
   No discount is given. The full amount is due within 30 days.
5. *n/EOM*—Read: net-end of month.
   No discount is given. The full amount is due by the end of the current month. If the invoice is dated after the 15th, the invoice is usually due at the end of the next month.

Invoice

An invoice shows information about the purchase of merchandise. Study the invoice in Illustration 9-1 so that you will be able to correctly identify the various parts.

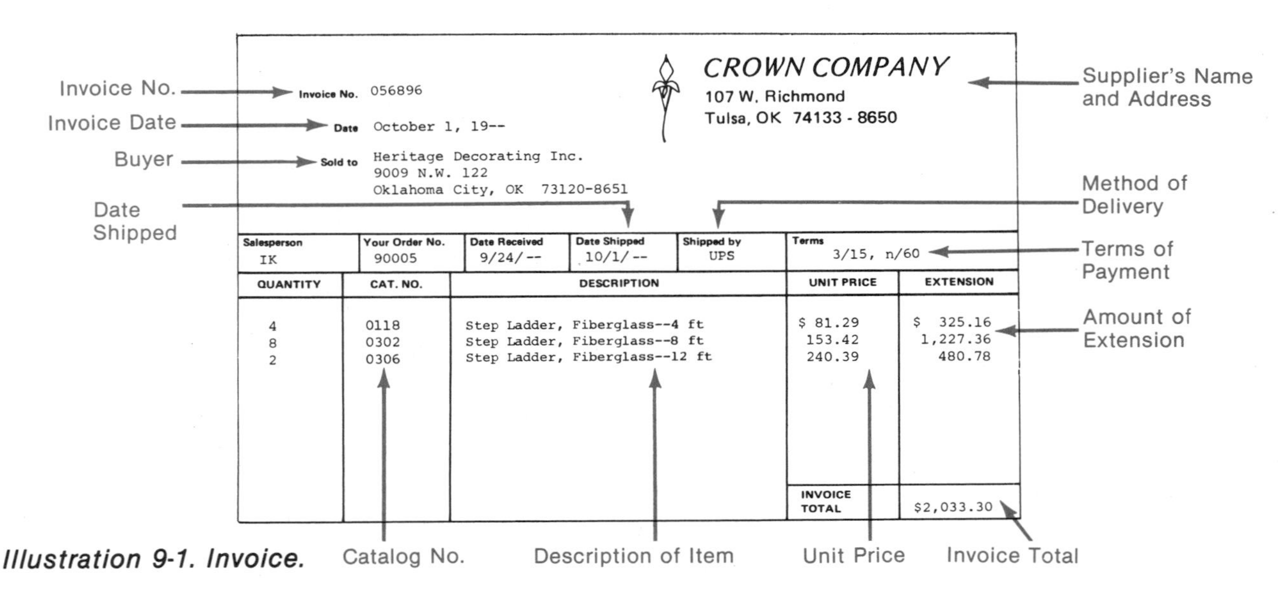

Invoice No. 056896

Date October 1, 19--

Sold to Heritage Decorating Inc.
9009 N.W. 122
Oklahoma City, OK 73120-8651

CROWN COMPANY
107 W. Richmond
Tulsa, OK 74133 - 8650

Salesperson	Your Order No.	Date Received	Date Shipped	Shipped by	Terms
IK	90005	9/24/--	10/1/--	UPS	3/15, n/60

QUANTITY	CAT. NO.	DESCRIPTION	UNIT PRICE	EXTENSION
4	0118	Step Ladder, Fiberglass--4 ft	$ 81.29	$ 325.16
8	0302	Step Ladder, Fiberglass--8 ft	153.42	1,227.36
2	0306	Step Ladder, Fiberglass--12 ft	240.39	480.78
			INVOICE TOTAL	$2,033.30

Illustration 9-1. Invoice.

Calculating Cash Discounts

Illustration 9-1 is dated October 1 with terms 3/15, n/60. The invoice must be paid within 15 days after October 1 to deduct a 3% discount from the invoice total.

The discount is figured using the percentage formula.

$$B \times R = P$$
$$\$2{,}033.30 \times 3\% = \$61.00$$

The **invoice price** is then calculated by subtracting the discount from the invoice total.

$$\$2{,}033.30 - \$61.00 = \$1{,}972.30$$

* ***Use the percent key to find the discount. Operate the minus key to obtain the invoice price or use the memory to determine the invoice price. If your calculator has a − % key, use it.***

Assume each of the following invoices is paid within the discount period. Find the cash discount and the invoice price.

Keep your decimal selector set on 2 unless instructed otherwise.

	Invoice Total	Terms	Cash Discount	Invoice Price
1.	$532.00	2/10,n/30	______	______
2.	$720.00	$2\frac{1}{2}\%$	______	______
3.	$2,344.65	3/15,n/30	______	______
4.	$175.87	3/10,n/15	______	______
5.	$316.95	2/15,n/30	______	______

Due Date

"Thirty days hath September, April, June, and November; all the rest have thirty-one, excepting February alone which has 28 in all, plus each leap year 29."

Richard Grafton

To determine the last day a cash discount can be taken or the *due date* of the invoice, add the number of days in the terms to the date of the invoice.

Example: An invoice is dated July 6 with terms 3/15, n/30.

1. To determine the last day the invoice can be paid to take the 3% discount, add 15 days to the invoice date: 6 + 15 = 21. The invoice must be paid by July 21 to get the 3% discount.
2. To find the last day the invoice can be paid (30 days):
   a. Find the number of days left in the month.

      31 days in July
      − 6 invoice date
      25 days left in July

b. Find the number of days into the next month to match the invoice terms.

$$\begin{array}{rl} 30 & \text{days—terms of invoice} \\ -25 & \text{days left in July} \\ \hline 5 & \text{days into August} \end{array}$$

The invoice is due August 5.

3. If the invoice terms were n/60, you would figure the due date as follows:

a.
$$\begin{array}{rl} 31 & \text{days in July} \\ -\ 6 & \text{invoice date} \\ \hline 25 & \text{days left in July} \end{array}$$

b.
$$\begin{array}{rl} 60 & \text{days—invoice terms} \\ -25 & \text{days in July} \\ \hline 35 & \text{days left on invoice terms} \\ -31 & \text{days in August} \\ \hline 4 & \text{days into September} \end{array}$$

The invoice is due September 4.

* ***This can be figured in one continuous operation on the calculator.***

Find the last date the largest cash discount can be deducted and the last day the invoice price of the bill is due for each of the following.

	Date of Invoice	Terms	Last Day of Maximum Discount	Last Day Invoice Due
1.	March 7	2/15,n/60	____________	____________
2.	Aug. 22	5/10,2/15,n/60	____________	____________
3.	Sept. 25	5/20,n/60	____________	____________
4.	May 3	5/15,n/30	____________	____________
5.	April 29	3/10,n/60	____________	____________

Practice

Find the discount due date, discount amount, and invoice price on the following invoices.

	Date of Invoice	Amount of Invoice	Terms	Date of Payment	Discount Due Date	Discount Amount	Invoice Price
1.	June 27	$740.00	3/15,n/30	July 9	________	________	________
2.	July 20	$617.00	2/15,n/30	Aug. 6	________	________	________
3.	Sept. 15	$839.00	2/10,1/30,n/60	Oct. 2	________	________	________
4.	May 4	$616.00	5/15,2/20,n/30	May 19	________	________	________
5.	Feb. 27	$143.00	3/10,2/15,n/30	March 10	________	________	________

Trade Discounts

A retailer buys goods from a wholesaler, manufacturer, or other supplier. A **price list** or catalog gives the "suggested retail price" of each item. This price is called the **list price**.

Rather than print a new catalog every time there is a price change, the seller gives a **trade discount** to the buyer. When the manufacturer's price changes, only the **discount rate** changes. The list price stays the same. A trade discount is deducted from the list price before the cash discount is figured.

Remember: Percentage = Base × Rate!

Example: A wholesaler's catalog may list Item B at $120, less a trade discount of 25%. The retailer will pay $90.

$120 × 25% = $30 (discount)
$120 – $30 = $90 (cost to retailer—net price)

The percent discount feature on the calculator can be used here.

Find the trade discount and the **net price** of the following items.

	Item	List Price	Discount Rate	Trade Discount	Net Price
1.	Flight Bag	\$50.50	12%	________	________
2.	Pullman Luggage	\$80.98	15%	________	________
3.	Carryon Luggage	\$45.97	20%	________	________
4.	Garment Bag	\$72.00	$33\frac{1}{3}$%	________	________
5.	Duffel Bag	\$55.60	$15\frac{1}{2}$%	________	________

Complements

Often the retailer needs to know only the net price. The amount of the discount is not needed. To find the net price, multiply the list price by the **complement** of the discount rate. To find a complement of the discount rate, subtract the discount rate from 100%.

Example: If a tape recorder's list price is \$75 less a 20% discount to retailers, then a retailer must pay 80% of the cost (100% − 20% = 80%). * ***The list price multiplied by the complement gives the net price.***

\$75 × 80% = \$60 (net price)

or

\$75 × 20% = \$15 (discount)
\$75 − \$15 = \$60 (net price)

* ***Can you see that this is a shortcut in finding the net price?***

Find the complement for each of the following.

	Trade Discount Rate	Complement		Trade Discount Rate	Complement
1.	30%	________	**2.**	$45\frac{1}{2}$%	________
3.	22%	________	**4.**	40%	________
5.	32%	________	**6.**	18%	________

Now, using the complement of the discount rate, find the net price in each of the following problems.

7. $85.50 less 20% = ________ **8.** $657.40 less 15% = ________

9. $87 less 22% = ________ **10.** $540 less 15.5% = ________

11. $45.95 less 4% = ________ **12.** $703.60 less 12% = ________

* *You can alternate between using the discount rate and its complement when working discount problems.*

Practice

Solve the following problems.

1. Find the discount to be deducted from a $80 list price if a 20% discount rate is allowed.

 ____

2. Find the net amount of an item listed for $125.50 if the discount is 30%.

 ____

3. Find the net amount of an item listed for $3,210 if the discount is 35%.

 ____

4. Find the discount to be deducted from a $99 item if a $33\frac{1}{3}$% discount rate is allowed.

 ____

5. Find the net amount of an item listed for $400 with a 25% discount.

 ____

Series Discounts

Retailers are offered different discounts depending on the volume of business and cost involved. One retailer might be offered a 40% discount while another retailer or a wholesaler is given a 35% discount plus an additional 10% discount. This **series discount** would be stated as 35, 10. It is not the same as a 45% discount. When there is a series discount, the second or third discount is allowed *after* the first discount has been deducted. The discounts must be computed in succession. They cannot be added.

Example: Find the cost of an item listed for $250 with trade discounts of 20, 10, and 5.

1.	Find the first discount:	$250 × 20% = $50
	Find the first net amount:	$250 – $50 = $200
2.	Find the second discount:	$200 × 10% = $20
	Find the second net amount:	$200 – $20 = $180
3.	Find the third discount:	$180 × 5% = $9
	Find the final net amount:	$180 – $9 = $171

Or you may use a *simpler method* using *complements*.

1. Find the complement of each discount rate.

   .8, .9, and .95

2. Multiply the complements by the price to find the net amount.

   $250 × .8 × .9 × .95 = $171

Using complements on the calculator simplifies computing net price.

* ***Remember, when you multiply by the complements, you find the net price. You do not find the discount. If you need to know the discount, the net price must be subtracted from the list price.***

Change each percent to a decimal and find its complement. Use the complement to find the net price.

Use multifactor multiplication.

Where should the decimal selector be set if you are doing a series of calculations?

1. $300.50 less 10-5-5 = ______________

2. $656.00 less 10-15-5 = ______________

3. $800.65 less 10-20-10 = ______________

4. $187.46 less 10-2$\frac{1}{2}$-5 = ______________

5. $736.15 less 15-5 = ______________

Single Discount Equivalent

If the same series of discounts is used frequently, it is useful to obtain the *single discount equivalent*. Compute the single discount equivalent by multiplying the complements together. Then subtract the complement product from 1.00 (100%) to get the single discount equivalent. Round to three decimal places.

Example: A 10-15-5 discount is offered on all items in a catalog.

.9 × .85 × .95 = .727

1.00 − .727 = .273 (single discount equivalent)

Use complements to find the single discount equivalent for each series discount. Round each single discount equivalent to at least three decimal places.

Use multioperations to find the single discount equivalent.

1. 20-20-5 = ______________ **2.** 20-10-10 = ______________

3. 5-10-5 = ______________ **4.** 10-10-5 = ______________

5. 15-5$\frac{1}{2}$-5 = ______________ **6.** 12-10-5 = ______________

Use the single discount equivalent of 10-20-5 to find the discount for each of the following problems. * ***Once the single discount equivalent is found, set the decimal selector on 2, and use constant multiplication to find the amount of the discount.***

	List Price	Discount		List Price	Discount
7.	$1,420.50	____________	**8.**	$271.00	____________
9.	$493.00	____________	**10.**	$630.60	____________

If only the net price is needed, use the product of the complements and do not subtract from 100%.

Example: 10-15-5 discount
$.9 \times .85 \times .95 = .727$
List Price $\times$ .727 = Net Price

Use the complement of the single discount equivalent of 15-20-5 to find the net price. * ***Once the single discount equivalent is found, set the decimal selector on 2, and use constant multiplication to find the net price.***

	List Price	Net Price		List Price	Net Price
11.	$350.00	____________	**12.**	$473.00	____________
13.	$295.00	____________	**14.**	$620.25	____________

Did you use constant multiplication?

EVALUATING YOUR SKILLS

The following problems evaluate your skills in the material covered in Units 1 through 8. Some of the problems will evaluate your ability to work problems mentally while others will evaluate your ability to use your calculator. Work the problems as quickly as you can.

1. Round $527.248 to the nearest cent. ____________
2. Round 243,455.09 to the nearest thousand. ____________
3. Convert $\frac{5}{9}$ to a decimal (to four places). ____________
4. Write $4\frac{3}{4}\%$ as a decimal. ____________
5. $2,673 \times 10 =$ ____________
6. $74,218 \div 100 =$ ____________
7. Estimate the product of $8,720 \times 31$. ____________
8. Write .36 as a fraction in lowest terms. ____________

9. 120 is what % of 600? ____________

10. What is 18% of 720? ____________

11. 93 miles is equal to how many kilometers? ____________

12. 284 is 20% of ? ____________

13. Multiply $\frac{3}{4} \times \frac{7}{15}$. ____________

14. $7.2 \times .044 =$ ____________

15. What is the cost of 5,420 items at $6 per C? ____________

APPLYING YOUR KNOWLEDGE

Using the information given, solve the following problems. Remember to use the correct calculator features.

1. What months have 30 days? ____________
2. What months have 31 days? ____________
3. What month has 28 or 29 days? ____________
4. Merchandise with a list price of $8,493 is sold with a discount of $397. What is the percent of discount allowed? ____________
5. The Ortiz Company can buy $7,800 of patio furniture from Firm A paying $110 freight charge with a 23-15 discount. Firm B offers the same merchandise for $8,200 with no delivery charges and a trade discount of 15-5-5.
   a. Should the Ortiz Company purchase the furniture from Firm A or Firm B? ____________

   b. How much is saved? ____________
6. Find the net price of merchandise listed at $345.78 with:
   a. a trade discount of 25-15. ____________

b. a single discount of 40%.

____

7. Watts Jewelry received a 35% trade discount amounting to $78. What was the list price of the merchandise? (Are you trying to find percentage, base, or rate?)

____

8. Find the net price on a $379 item at 35-10 discount.

____

9. Find the discount amount on a $359 item with a 20-10-2$\frac{1}{2}$ discount.

____

10. Which will give the lower net price on a $235 item: a 25-15 discount or a 15-10-5 discount?

____

11. Compute the single discount equivalent of a 35-20-15 discount.

____

12. An architectural firm needs to purchase a new drafting table. Complete the following cost comparison chart to determine which company offers the lowest cost after all discounts have been taken.

	Company	Unit Price	Trade Discount Rate	Net Price	Cash Discount Terms	Invoice Price
a.	A H Co.	$625.00	20-10	________	2/10,n/30	________
b.	Xarre Co.	$620.15	5-10-5	________	3/10,n/30	________
c.	Ming Co.	$640.75	10-15-5	________	3/10,1/15,n/30	________

ADVANCED APPLICATIONS

Using the information given, solve the following problems.

1. Find the net amount of an invoice for $8,756 including $250 defective merchandise which must be returned, if terms are 4/15, n/60 and payment is made within the discount period.

____

2. An invoice of $4,987 includes a $57 prepaid freight charge and $103 spoiled merchandise. Terms of 2/10,1/15,n/30 are dated January 4. If payment is made on January 18, how much is the net amount due? * *Discount is not calculated on freight.*

____

3. The following checks were received in payment of invoices for merchandise purchased. The terms of the invoices are 4/10, 2/15, n/60. Determine if the correct discount was taken and record any errors. Use parentheses to indicate any overpayments.

	Date Check Received	Invoice Price	Date of Invoice	Check Amount	Errors
a.	April 3	$364.00	March 23	$348.44	______
b.	March 17	$354.00	March 3	$339.14	______
c.	Oct. 2	$3,486.00	Sept. 24	$3,346.56	______
d.	Aug. 16	$791.00	August 6	$775.18	______

4. An invoice listed a cash discount of 4%. The amount of the discount is $89. What is the amount due?

____

5. The Xerrie Company imports teak from East India. On November 16, the company received a shipment with an invoice in the amount of $7,985, which included a trade discount of 15-22. The invoice was dated November 22 and had terms of 3/10, 2/15, n/60. What amount will be paid by Xerrie Company if the check is dated December 1?

____

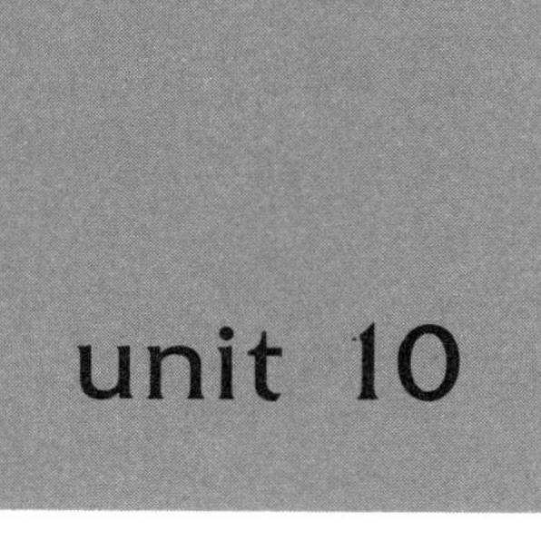

unit 10 Banking Deposits and Checks

"A habit is like a soft bed—easy to get into but hard to get out of."

The modern banking system provides many useful services to individuals and businesses. Banks provide a safe place to keep money, a means to make payments with convenience, and help with many financial aspects of business operations. Therefore, it is important that you know and understand how to use the services provided by banks.

After studying this unit, you should be able to:

1. open a checking account
2. write a check
3. make a deposit

MATH TERMS AND CONCEPTS

The following terminology is used in this unit. Become familiar with the meaning of each term so that you understand its usage as you develop math skills.

1. **ABA (American Bankers Association) Numbers**—identification numbers that are assigned to each bank.
2. **Average Balance**—the beginning balance plus deposits and minus withrawals divided by the number of days in the month.
3. **Check**—an order with an authorized signature by the owner of an account directing a bank to make a payment from the account.
4. **Check Register**—an individual's record of checks written, deposits made, charges made by the bank, and interest earned.
5. **Check Stub**—the part of the check that stays in the checkbook and contains information about the checks written and deposits made.
6. **Checkbook Balance**—the current checking account balance recorded in the check register.
7. **Checking Account**—money deposited in a bank account on which checks are written.
8. **Deposit**—money added to an account.
9. **Depositor**—the owner of the account.
10. **Deposit Slip**—a form used to list all cash and checks the depositor will enter into an account.

11. **Endorsement**—a signature on the back of a check allowing the check to change hands.
12. **Maintenance Fee**—(see service charge).
13. **MICR (Magnetic Ink Character Recognition)**—machine readable numbers printed along the bottom of a check that include numbers assigned to both the depositor and the bank.
14. **Minimum Balance**—a certain amount that must be maintained in a checking account at all times to avoid fees.
15. **Money Market Account**—an account that earns a variable interest rate and that requires a minimum balance be maintained.
16. **Money Market Rate**—an interest rate based on the rate most financial institutes pay.
17. **NOW Account**—a checking account on which the bank pays interest.
18. **Payee**—the person or business to whom a check is written.
19. **Payor**—the person who signs a check.
20. **Service Charge**—a charge made by the bank for handling an account.
21. **Signature Card**—a form showing the depositor's authorized signature.
22. **Withdrawal**—money taken from an account by the owner.

Checking Accounts

Checking accounts provide a safe place to keep money and are convenient and easy to use. To open a checking account, you must visit a bank and complete a **signature card** to be kept on file at the bank. The signature card should be signed the same way all checks will be signed. You will then fill out a **deposit slip** showing the amount of money used to open your account. The bank will provide checks and deposit slips with your name and account number printed on them. Your account number is printed on the checks in magnetic ink (**MICR**) so computers can sort and process the checks.

There are different types of checking accounts designed to fit individual needs. Shop around to determine the best checking account for your needs. Most banks will offer several types of checking accounts.

Example: Bank XYZ offers four types of checking accounts:

A. Regular Checking
- —$200 minimum to open account
- —$200 **average balance**—no **service charge**
- —Average balance below $200—$7.50 service charge
- —Unlimited check writing privileges

B. Special Account
- —No minimum to open
- —$2.00 **maintenance fee**
- —First 7 checks—no service charge
- —After 7 checks—$.25 per check

C. **NOW Account**
- —$500 minimum to open account
- —$8.50 service charge if average balance falls below $500
- —Interest is earned at a rate of $5\frac{1}{4}$%; no interest is earned if average balance falls below $500
- —Unlimited check writing privileges

D. **Money Market Account**
—$1,000 minimum to open account
—Interest earned is based on **money market rate**
—$10 service charge if **minimum balance** falls below $1,000
—No interest is earned if minimum balance falls below $1,000
—Unlimited check writing privileges

Determine the type of account that would be best for each of the following customers.

1. A student who writes about 6 checks per month and has an average balance of about $250.

____

2. A married couple who writes about 25 checks per month and has an average balance of about $600.

____

3. An individual who writes 3 checks each month, has an average balance of well over $1,000, and wants to earn a rate of interest based on money market rate conditions.

____

Deposit Slips

Each time a **deposit** is made to an account, a deposit slip must be filled out. The currency (dollar bills), coins, and checks are listed separately and then totaled. (See Illustration 10-1.) Checks are listed by the **ABA number** or by the name of the **payor**. (See Illustration 10-1). Businesses often list checks by the payor's name and use a copy of the deposit slip as an additional record of payments. If any cash is to be given to the **depositor**, the amount is deducted and a new total obtained.

Account No. 066-4163

Date 9/12/--	DOLLARS	CENTS
CURRENCY	16	00
COIN	5	73
CHECKS (List each separately)		
1 Joe Cheng	27	42
2 42-44	79	50
3 39-88	102	45
4 D. R. Larsten	83	98
5		
6		
7		
8		
SUBTOTAL	315	08
LESS CASH RECEIVED	40	00
TOTAL	275	08

Illustration 10-1. Deposit Slip.

Follow these steps to complete a deposit slip:
1. Record the date—using today's date.
2. Total and record the currency.
3. Total the coins and record.

4. List each check separately by ABA number or payor.
   * *Use the back of the deposit slip to record additional checks.*
5. Total the deposit slip. Bring the back total to the front.
6. Subtract any cash received.
7. Total the deposit slip.

Complete a deposit slip for each of the following.

Use GT/Memory to help you complete the deposit slip.

	Coins	Bills	Checks	
1.	Three quarters	Four 20's	#40-3291	$54.85
	Four dimes	Three 10's	#64-6643	$129.00
	Seven nickels	Five 5's	#51-287	$75.83
	Nine pennies	Eleven 1's	#169-0058	$60.40
2.	One half-dollar	Two 20's	#28-9940	$38.97
	Six quarters	Seven 10's	#201-67	$261.05
	Ten dimes	Six 5's	#148-6893	$130.00
	Two nickels	Sixteen 1's	#48-57	$46.95
	Eight pennies	Cash received, $50		

Account No. 001-5264

Date	DOLLARS	CENTS
CURRENCY		
COIN		
CHECKS (List each separately) 1		
2		
3		
4		
5		
6		
7		
SUBTOTAL		
LESS CASH RECEIVED		
TOTAL		

Account No. 370-1048

Date	DOLLARS	CENTS
CURRENCY		
COIN		
CHECKS (List each separately) 1		
2		
3		
4		
5		
6		
7		
SUBTOTAL		
LESS CASH RECEIVED		
TOTAL		

Checks and Check Registers

Each time you write a **check** you are directing the bank to deduct money from your account. Study Illustration 10-2 to be able to correctly identify the parts of a check and **check stub**.

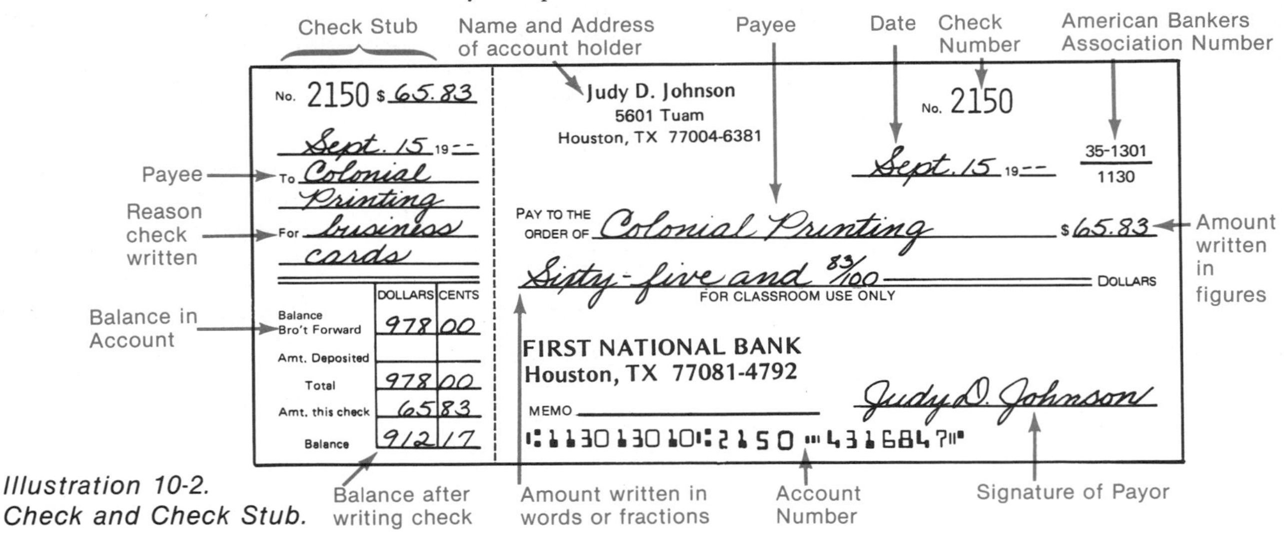

Illustration 10-2. Check and Check Stub.

Some individuals purchase checks with a **check register** rather than check stubs. The same information is recorded in the register as on a check stub. It is important that the **checkbook balance** be kept up to date and a new balance recorded each time a check is written or a deposit made.

The check and stub shown in Illustration 10-2 have been filled in correctly using these guidelines.

1. Use a typewriter or ink pen to write checks. Do not use a pencil.
2. Write the date and **payee**.
3. Write the amount close to the dollar sign so that the amount cannot be changed.
4. Write the amount of the check in words. If there is a difference, the amount written in words is paid.
   a. Capitalize the first word only.
   b. Hyphenate all numbers from twenty-one through ninety-nine.
   c. Omit writing the word dollars since the word appears at the end of the line.
   d. Use the entire space provided for the amount. Begin at the extreme left, spelling out the amount. Express cents as a fraction of 100. For example, $1,762.89 would be written: One thousand seven hundred sixty-two and 89/100 dollars.
   e. A check for an amount less then a dollar, for example $.49, would be written: Only forty-nine cents.

   * ***Try to avoid writing checks for less than one dollar.***
5. If an error is made, do not attempt to correct it. Write the word VOID across the check and the check stub. Then write a new check.
6. Sign the check the same way the signature card was signed.
7. Fill in the check or check register completely. Be sure you always determine a new balance.

   * ***Keep your checkbook up to date!***

Solve the following problems.

1. Write the following amounts as you would on a check.

a. $27.48 ____________________

b. $261.50 ____________________

c. $.83 ____________________

d. $1,420.75 ____________________

e. $91.62 ____________________

2. The balance on check stub #348 is $1,470.84. A deposit of $360 is made. Check #349 is written in the amount of $53.62. Find the new balance.

____

3. Write the correct balance after each deposit or check shown below.

	June 7:	Balance—		$620.48
a.	9:	Check —$127.14	**a.**	________
b.	10:	Deposit—$310.00	**b.**	________
c.	12:	Check — $28.96	**c.**	________
d.	14:	Check — $74.53	**d.**	________
e.	17:	Check —$129.40	**e.**	________
f.	20:	Check — $61.25	**f.**	________

4. Complete the following check and check stub. The balance brought forward is $671. Write the check for $7.41 using the current date and your name. Make the check payable to Klinger Electric for repairs.

No. 2175 $______

______ 19____

To ______

______

For ______

______

	DOLLARS	CENTS
Balance Bro't Forward		
Amt. Deposited		
Total		
Amt. this check		
Balance		

No. 2175

73-101 / 2421

______ 19____

PAY TO THE ORDER OF ______ $______

______ DOLLARS

FOR CLASSROOM USE ONLY

MUNY BANK
Covington, KY 41011-4798

FOR ______ ______

⑆242101013⑆ 2175 ⑈ 5896134⑈

5. The checkbook balance on March 1 was $341.79. During the month deposits were made in the amounts of $55.40, $132.48, and $210.30. Checks were written for $76.39, $283.21, $53.70, $14.32, $8.47, $149.60, $41.85, $66.00, and $24.56. What was the balance on March 31?

______

Endorsing a Check

A check must be endorsed before it can be cashed. The Funds Availability Act, which regulates the way checks are processed through the Federal Reserve System, places strict regulations on where endorsements may be placed. An **endorsement** must be made within $1\frac{1}{2}$ inches of the trailing left edge of the back of the check. If anything is written outside of the $1\frac{1}{2}$ inch limit, it may cause the check to be returned. The endorsement can be one of three types.

1. *Blank Endorsement*: The signature of the payee is written exactly as it is written on the front of the check. Anyone having possession of the check with a blank endorsement may cash it.

   Example: *Sue Baker*

2. *Restrictive Endorsement*: A restrictive endorsement limits who may receive a check. For example, a restrictive endorsement may have "For deposit only" above the payee's signature. This means

the check can only be deposited in the payee's account. Using a restrictive endorsement allows one to mail a check safely.

Example:

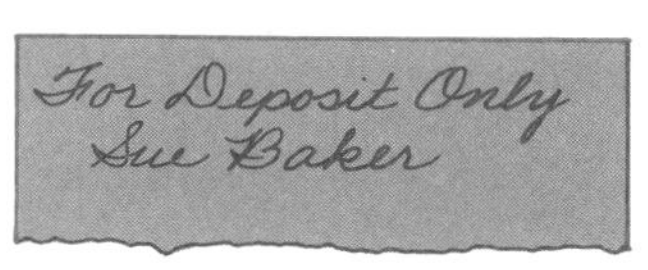

3. *Special Endorsement:* A special endorsement makes a check payable to a third party. Only the third party may cash the check.

Example:

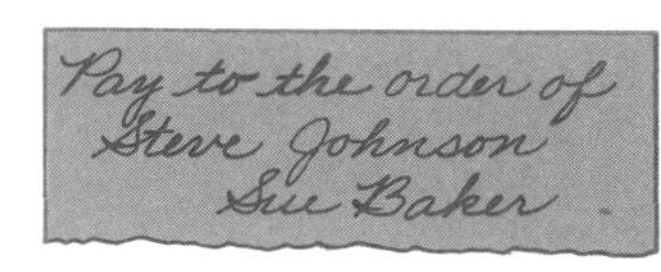

Using your name, correctly endorse the checks given below. The type of endorsement is indicated below each check. Make the special endorsement payable to Les Newson.

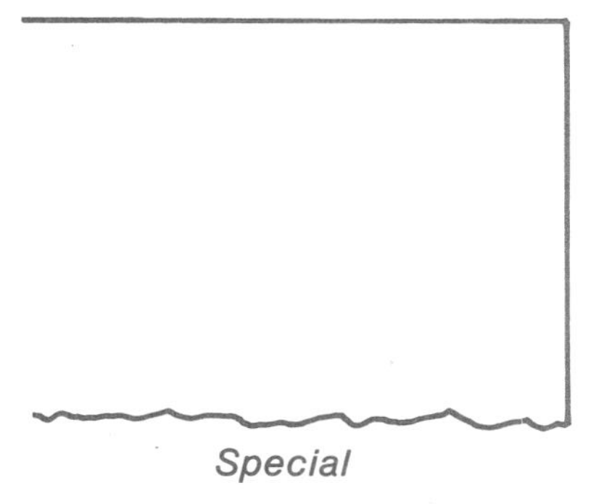

Special

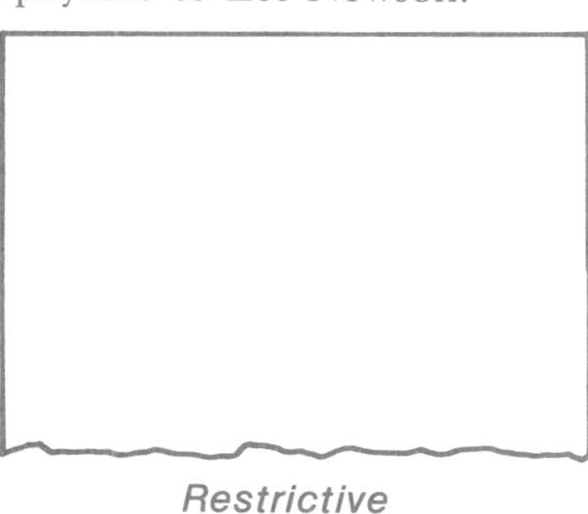

Restrictive

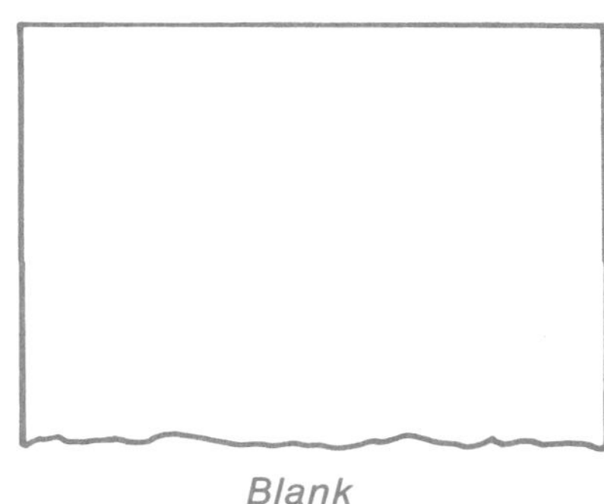

Blank

EVALUATING YOUR SKILLS

The following problems evaluate your skills in the material covered in Units 1 through 9. Some of the problems will evaluate your ability to work problems mentally while others will evaluate your ability to use your calculator. Work the problems as quickly as you can.

* *Remember to round all answers to two decimal places unless otherwise instructed.*

1. Reduce $3\frac{8}{12}$ to lowest terms. ____________

2. Change $\frac{23}{7}$ to a mixed number. ____________

3. Add $4\frac{2}{3} + 5\frac{1}{5}$. ____________

4. A person's height is 128 cm. Convert this to feet and inches. ____________

5. Write $\frac{3}{4}$% as a decimal. ____________

6. Change the fraction $\frac{20}{35}$ to a percent. ____________

7. Which is larger, $\frac{5}{8}$ or $\frac{2}{3}$, and by how much? ______

8. Estimate the product of $57,312 × 9.5%. ______

9. Convert the following fractions to decimals and add: $\frac{7}{9}$, $\frac{12}{17}$, $\frac{2}{7}$, $\frac{7}{15}$
   * *Carry out decimals to four places.* ______

10. What part of 80 is 64?
    a. expressed as a fraction ______

    b. expressed as a percent ______

11. 2 is what % of 1? ______

12. Solve: (28.4 + 25.89) × 44 = ______

13. The United States published a list of their largest trading partners in Western Europe. The exports, in billions of dollars, amounted to the following: United Kingdom, $33.7; Germany, $9.3; France, $7.1; and Netherlands, $8.6. What was the average of U.S. exports to Western Europe? ______

14. An item cost $.98 for 14 oz. How much would it cost per pound? ______

15. A company is offered a trade discount of 4% and terms of 2/10, n/30 on an invoice totaling $2,619.20. What amount is due if it is paid within 10 days of the date of the invoice? ______

APPLYING YOUR KNOWLEDGE

Solve the following problems using the information given.

1. A bank charges a $2.00 maintenance fee per month plus $.20 for every check after the first five checks written. What would the total charges be if the following number of checks were written?
   a. 4 checks written

   ______

b. 12 checks written

____

2. Shawn Dillard wrote a check for $2,344.62 to pay for office equipment. In completing the check, how should the amount be written in words?

____

3. Kyle Corbin manages a golf shop. During the month of March he made four deposits: $375.16, $521.20, $362.57, $294.00. He wrote 11 checks: $172.39, $62.84, $137.86, $15.25, $55.38, $275.40, $47.18, $70.00, $70.00, $324.90, $250.65. If he had a beginning balance of $460.28, find his ending balance.

____

4. Linda Rauch mails her deposits to her bank. How should she endorse her checks to protect her deposit?

____

5. The Ski Shop deposited the following:

Four 20's	Ten quarters
Two 10's	Nine dimes
Four 5's	Twelve nickels
Seventeen 1's	Two pennies

Checks: $31.24, $207.84, $135.25, $18.90, $10.70, $2.17, $37.76
Find the total deposit.

____

6. Using the following information, complete the check register for Arnold's Candy Shop:

Feb. 15—Check #277 to Candy Strip, Inc. for Invoice #435 in the amount of $87.50
15—Check #278 to Sun Newspaper for advertising expense in the amount of $220.00
18—Deposit made in the amount of $720.80
19—Check #279 to Calhoun's Supply Co. for purchases in the amount of $208.64

PLEASE BE SURE TO **DEDUCT** CHARGES THAT AFFECT YOUR ACCOUNT

NO.	DATE	DESCRIPTION OF TRANSACTION	SUBTRACTIONS: AMOUNT OF PAYMENT OR WITHDRAWAL (−)	✓	OTHER	ADDITIONS: AMOUNT OF DEPOSIT OR INTEREST (+)	BALANCE FORWARD
							507 46
276	2/12	TO Polk Sugar Co. FOR Invoice #0132	126 43				BAL 381 03
	2/14	TO FOR				279 48	BAL 660 51
		TO FOR					BAL
		TO FOR					BAL
		TO FOR					BAL
		TO FOR					BAL

ADVANCED APPLICATION

Richard Amman owns a hardware store. He keeps careful records of cash payments by check in his check register. Use the blank deposit slips, checks, and the check register on pages 197-198 to record the following transactions.

Record Richard's balance brought forward of $2,320.37 on the check register.

June 22—Paid B & R Majors for Invoice #0431 in the amount of $79.80. Write Check #170.

June 23—Made a deposit in the amount of $442.76, ABA #352-26. Fill out deposit slip for Account #4116108.

June 23—Paid Adlerman Co. for merchandise on Invoice #012372. Amount is $478; took a 10% discount. Write Check #171.
* *Deduct 10%.*

Complete the remaining transactions using only the check register. Be sure to record a new balance for each check and deposit.

June 25—Paid $350.25 rent to Halverson Agency, Check #172.

June 26—Bought store supplies from Whittmore Office Supply. Wrote Check #173 for $42.70.

June 30—Paid insurance premium to Chung Ins. Co. Wrote Check #174 for $124.15.

June 30—Paid $314.20 on note to First Bank, Corp., Check #175.

June 30—Wrote Check #176 to Axis Co. for repairs. Amount of check—$179.16.

July 1 —Made deposit in the amount of $525.80.

July 2 —Paid Invoice #202 to Scutter Co. for merchandise. Wrote Check #177 in the amount of $139.40.

July 3 —Wrote Check #178 to R. & R. Utility Company for utility expense in the amount of $121.72.

July 6 —Wrote Check #179 to S. C. Telephone for long distance bill in the amount of $42.17.

July 6 —Made deposit in the amount of $618.40.

July 7 —Paid Sam Holden's salary of $404.78, Check #180.

July 7 —Paid Ryan Rutz's salary of $428.38, Check #181.

July 7 —Paid Karen O'Connell's salary of $443.40, Check #182.

July 8 —Bought store supplies from Whittmore Office Supply. Wrote Check #183 for $17.80.

July 9 —Paid Invoice #2178 with Check #184 to Herwig Distributors for merchandise. Amount of invoice is $186.40. Took a 3% discount.
* *Deduct 3%.*

July 9 —Made deposit in the amount of $793.07.

AMMAN'S HARDWARE
1700 MAIN STREET
CEDAR FALLS, IA 50613-3512

170

________________ 19_____ 72-181/739

PAY TO THE ORDER OF ______________________________ $__________

____ DOLLARS

FOR CLASSROOM USE ONLY

FIRST NATIONAL BANK
CEDAR FALLS, IA 50613-0021

MEMO________________________ ______________________________

⑆073901819⑆0170⑈4116108

AMMAN'S HARDWARE
1700 MAIN STREET
CEDAR FALLS, IA 50613-3512

171

________________ 19_____ 72-181/739

PAY TO THE ORDER OF ______________________________ $__________

____ DOLLARS

FOR CLASSROOM USE ONLY

FIRST NATIONAL BANK
CEDAR FALLS, IA 50613-0021

MEMO________________________ ______________________________

⑆073901819⑆0171⑈4116108

Account No. 4116108

Date __________	DOLLARS	CENTS
CURRENCY		
COIN		
CHECKS (List each separately) 1		
2		
3		
4		
5		
6		
7		
SUBTOTAL		
LESS CASH RECEIVED		
TOTAL		

PLEASE BE SURE TO **DEDUCT** CHARGES THAT AFFECT YOUR ACCOUNT

NO.	DATE	DESCRIPTION OF TRANSACTION	SUBTRACTIONS AMOUNT OF PAYMENT OR WITHDRAWAL (−)		✓	OTHER	ADDITIONS AMOUNT OF DEPOSIT OR INTEREST (+)		BALANCE FORWARD	
		TO								
		FOR							BAL	
		TO								
		FOR							BAL	
		TO								
		FOR							BAL	
		TO								
		FOR							BAL	
		TO								
		FOR							BAL	
		TO								
		FOR							BAL	
		TO								
		FOR							BAL	
		TO								
		FOR							BAL	
		TO								
		FOR							BAL	
		TO								
		FOR							BAL	
		TO								
		FOR							BAL	
		TO								
		FOR							BAL	
		TO								
		FOR							BAL	
		TO								
		FOR							BAL	
		TO								
		FOR							BAL	
		TO								
		FOR							BAL	
		TO								
		FOR							BAL	
		TO								
		FOR							BAL	
		TO								
		FOR							BAL	
		TO								
		FOR							BAL	

unit 11

Bank Reconciliations and Services

"Only your best behavior is good enough for daily use."

Comparing your check register with the bank's records is necessary to maintain an accurate balance. It is important that you reconcile your checkbook and bank statement each month. You should also be aware of all the services offered by banks.

After studying this unit, you should be able to:

1. reconcile a bank statement and checkbook
2. compare services provided by banks
3. compute charges to retailers for use of bank credit cards

MATH TERMS AND CONCEPTS

The following terminology is used in this unit. Become familiar with the meaning of each term so that you understand its usage as you develop math skills.

1. **Bank Credit Card**—a service to businesses allowing them to offer credit to customers (Visa, Mastercard).
2. **Bank Draft**—authorization to have a bank make automatic payments.
3. **Bank Reconciliation**—the process of bringing the bank statement and checkbook balance into agreement.
4. **Bank Statement**—the bank's record of a depositor's account showing deposits, withdrawals, bank charges, and interest.
5. **Canceled Check**—a check processed by the bank.
6. **Electronic Transfer**—the process of moving funds from one account to another without writing a check or completing a deposit slip.
7. **Insufficient Check**—a check processed by the bank for which there is not enough money in the account to cover it.
8. **Outstanding Check**—a check that has been written but has not been processed by the bank on the date the bank statement is prepared.
9. **Outstanding Deposit**—a deposit made but not shown on the bank statement.

A monthly **bank statement** is sent to each depositor showing all checks and deposits processed by the bank. (See Illustration 11-1.) It is the bank's record of the depositor's checking account. Since the checkbook balance usually does not agree with the bank statement balance, the balances must be reconciled each month. **Reconciling** brings the checkbook balance and bank statement balance into agreement—that is, the same figure. **Bank reconciliation** is the best way to know exactly how much money is in an account.

BANK STATEMENT

Previous Balance	No.	Check Amount	Deposit Amount	Service Charge	New Balance
$1,667.43	220	124.79			1,542.64
	221	275.30			1,267.34
	224	86.92			1,180.42
	226	418.00	275.00		1,037.42
	227	306.50			730.92
				3.50	727.42

Illustration 11-1. Bank Statement.

The checkbook balance and bank statement balance may not agree because a service charge or maintanence fee may have been deducted on the bank statement. Also, checks written and amounts deposited may not yet have been processed by the bank. These are called **outstanding checks** and **outstanding deposits**.

Most bank statements have a form on the back to help you bring the two balances into agreement. It may look something like the form in Illustration 11-2.

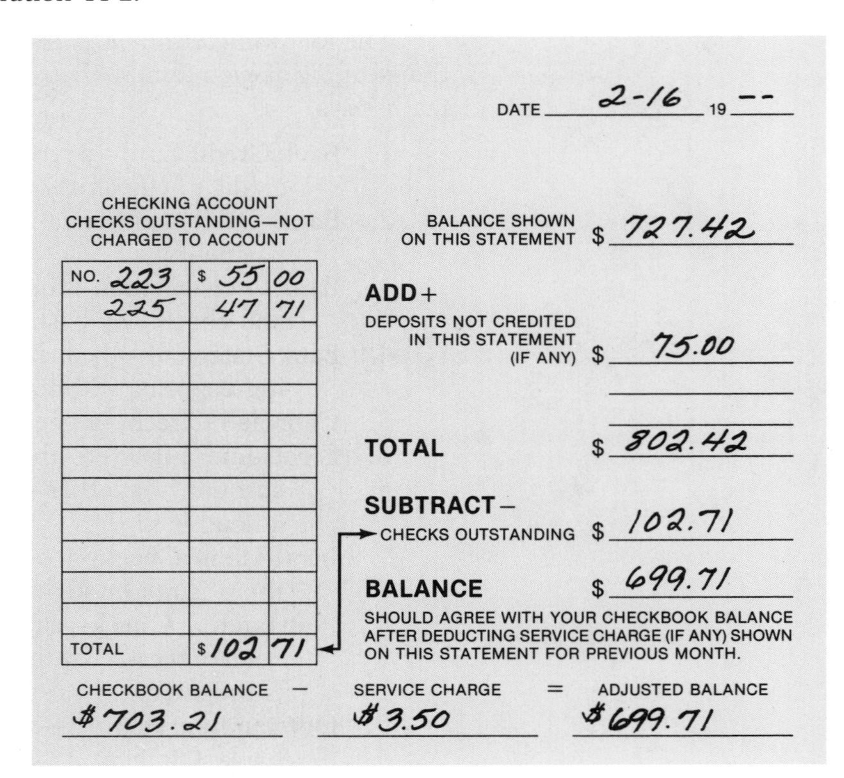
DATE 2-16 19 --

CHECKING ACCOUNT CHECKS OUTSTANDING—NOT CHARGED TO ACCOUNT

NO.	$	
223	55	00
225	47	71
TOTAL	$102	71

BALANCE SHOWN ON THIS STATEMENT $ 727.42

ADD+

DEPOSITS NOT CREDITED IN THIS STATEMENT (IF ANY) $ 75.00

TOTAL $ 802.42

SUBTRACT–

CHECKS OUTSTANDING $ 102.71

BALANCE $ 699.71

SHOULD AGREE WITH YOUR CHECKBOOK BALANCE AFTER DEDUCTING SERVICE CHARGE (IF ANY) SHOWN ON THIS STATEMENT FOR PREVIOUS MONTH.

CHECKBOOK BALANCE	–	SERVICE CHARGE	=	ADJUSTED BALANCE
$703.21		$3.50		$699.71

Illustration 11-2. Bank Statement Reconciliation Form.

You must keep a record of all checks written and all deposits made to reconcile your checkbook and bank statement.

Follow these steps in reconciling a checkbook and bank statement:

1. Arrange the **canceled checks** in numerical order. Put a check mark on each check stub or in the check register for each check returned. Any check not marked off is outstanding.
   Some banks do not return the cancelled checks but list each check's number and amount on the statement. Follow the same procedure and check off each check listed on the bank statement using the check register.
2. Check off all deposits shown on the bank statement in the check stub or in the check register. Add any outstanding deposits to the bank statement balance.
3. List all outstanding checks and subtract from the balance on the bank statement.
4. Subtract any charges or fees from your checkbook balance.
5. Compare the adjusted bank statement balance with the adjusted checkbook balance. If they are equal, you have reconciled your checkbook and bank statement.

* *You will learn later what to do if the two balances do not agree.*

Solve the following problems.

1. The following list shows the adjustments necessary in bank reconciliation. Write the word add or subtract under the appropriate column.

	Bank Statement Balance	Checkbook Balance
a. Service Charge	________	________
b. Outstanding Deposits	________	________
c. Outstanding Checks	________	________

2. Complete the following bank statement reconciliation form using this information:
Bank statement balance on November 30: $983.16
Checkbook balance: $928.96
Outstanding Checks: #177, $56.60; #178, $27.43; #180, $49.50
Outstanding Deposit: $71.83
Service Charge: $7.50

DATE ________ 19 ____

CHECKING ACCOUNT
CHECKS OUTSTANDING—NOT CHARGED TO ACCOUNT

NO.	$	
TOTAL	$	

BALANCE SHOWN ON THIS STATEMENT $ ________

ADD +
DEPOSITS NOT CREDITED IN THIS STATEMENT (IF ANY) $ ________

TOTAL $ ________

SUBTRACT –
CHECKS OUTSTANDING $ ________

BALANCE $ ________
SHOULD AGREE WITH YOUR CHECKBOOK BALANCE AFTER DEDUCTING SERVICE CHARGE (IF ANY) SHOWN ON THIS STATEMENT FOR PREVIOUS MONTH.

CHECKBOOK BALANCE — SERVICE CHARGE = ADJUSTED BALANCE

________ ________ ________

T Forms

If the bank does not provide a reconciliation form, you can prepare your own reconciliation statement using a T form. (See Illustration 11-3.)

BANK RECONCILIATION
April 1, 19--

Bank Statement Balance	$678.22	Checkbook Balance	$710.37
Add Outstanding Deposit	+100.00		
	$778.22		
Deduct Outstanding Checks:		Deduct Service Charge	− 4.00
#315 $52.84			
#316 +19.01			
	− 71.85		
Adjusted Bank Statement Balance	$706.37	Adjusted Checkbook Balance	$706.37

Illustration 11-3. Reconciliation T Form.

Prepare a bank reconciliation for the following problem.

1. Bank statement balance on July 1: $2,320.72
Checkbook balance: $2,245.05
Outstanding deposit: $460.00
Service charge: $7.50
Outstanding checks: #546, $136.75; #549, $157.32; #550, $200.16; #552, $48.94.

BANK RECONCILIATION
July 1, 19--

Finding Errors

After going through the reconciliation steps, the adjusted checkbook balance may not agree with the adjusted bank statement balance. Check for errors in these areas:

1. *Deposits*—Verify the deposits on the bank statement with the checkbook record. If a deposit was not added into your check register, add it to your checkbook balance. If a deposit was recorded twice in your check register, subtract it from the checkbook balance.

2. *Checks*—See if all the checks were recorded in the check register. Match the checks listed on the bank statement with the ones recorded in the check register. If a check was not recorded, subtract it from the checkbook balance.
3. *Addition/Subtraction Errors*—Verify the accuracy of all the addition and subtraction used to find a new checkbook balance after each entry.

   * ***Add or subtract the amount of any error at the end of your check register rather than attempting to change each amount.***
4. *Errors in Recording*—Verify that all checks were recorded accurately. A common error is to transpose figures when recording checks. If a recording error is made, find the difference and add or subtract the difference in the checkbook balance.
5. *Outstanding Checks*—Subtract any outstanding checks from the bank statement balance.
6. *Service Charge*—Subtract the service charge from the checkbook balance.
7. *Extra Charges*—Subtract any extra charges from the checkbook balance. These charges will be shown on the bank statement as insufficient fund charges, check charges, **bank draft charges**, etc.
8. *Transfer to Savings*—Subtract from the checkbook balance the amount of any money transferred from your checking account to your savings account.

* ***Ask yourself who doesn't know about a transaction—bank or checkbook. If the bank doesn't know, then adjust the bank statement. If you don't know, then adjust the checkbook.***

Solve the following problems.

1. The following list shows the adjustments necessary in bank reconciliation. Write the word add or subtract under the appropriate column.

	Bank Statement Balance	Checkbook Balance
a. Check omitted on stub	________	________
b. Deposit recorded twice	________	________
c. Deposit mailed too late	________	________
d. Deposit omitted on checkbook	________	________
e. Bank draft	________	________
f. Check for $21.40 entered as $24.10	________	________
g. Check written for $15.60 recorded on stub as $15.00	________	________

2. Use a T form to complete the following bank reconciliation on November 1.
Checkbook balance: $556.41
Bank statement balance: $682.14
No service charge
Outstanding checks: #516, $17.62; #519, $44.68; #520, $54.73; #522, $54.20
Deposit recorded twice in checkbook: $45
Check for $35.50 was recorded as $35.00

BANK RECONCILIATION
November 1, 19--

3. The following check register and bank statement information are given. Use a T form to complete the bank reconciliation on June 1. Use the blank lines in the check register to record corrections to the checkbook balance. Put a check mark beside the amount of each check and deposit processed by the bank.

   Bank statement balance: $1,388.30
   Electronic transfer to insurance company not deducted from checkbook: $86.40
   Service charge: $8.75
   Outstanding checks: Determine from the checks processed by the bank.
   Outstanding deposit: $136
   Charge on bank statement for personalized checks: $14.50
   Checks processed by bank: #423, #425, #426, #428
   Check #428 processed by bank for $92.64

PLEASE BE SURE TO **DEDUCT** CHARGES THAT AFFECT YOUR ACCOUNT			SUBTRACTIONS				ADDITIONS		BALANCE FORWARD	
NO.	DATE	DESCRIPTION OF TRANSACTION	AMOUNT OF PAYMENT OR WITHDRAWAL (−)		✓	OTHER	AMOUNT OF DEPOSIT OR INTEREST (+)		1,408	34
423	5/10	TO S. W. Utility Co. FOR Electricity	65	10					BAL 1,343	24
424	5/13	TO Holmes, Inc. FOR Rent	345	00					BAL 998	24
425	5/18	TO Trent Dickens FOR Delivery Fee	12	00					BAL 986	24
426	5/19	TO Brinker Bros. FOR 4-29 bill	32	95					BAL 953	29
	5/19	TO FOR					216	50	BAL 1,169	79
427	5/20	TO Petty Cash FOR	27	40					BAL 1,142	39
428	5/23	TO Lee, Inc. FOR Office Supply	82	64					BAL 1,059	75
	5/24	TO FOR					75	80	BAL 1,135	55
429	5/26	TO Perkins & Co. FOR Insurance Premium	102	60					BAL 1,032	95
	5/27	TO FOR					136	00	BAL 1,168	95
		TO FOR							BAL	
		TO FOR							BAL	
		TO FOR							BAL	
		TO FOR							BAL	

BANK RECONCILIATION
June 1, 19--

Bank Credit Cards

Another service provided to retailers is the **bank credit card**. Customers may charge purchases of goods and services at businesses that accept these cards. The business pays a fee for allowing customers to use the cards. The fee is based on the total charge card sales for the month. The fee may be 1% to 6%. The business shows the charge sales on the deposit slip. The bank deposits the total amount in the business' account.

Example: If a business has bank card charges of $1,763 and pays a fee of 6%, the charge would be calculated as follows:

$1,763 × .06 = $105.78

The amount of the fee is subtracted from the business' checking account. The business receives a form each month showing the amount of the charge. When the business reconciles the checkbook and bank statement, the bank card fee is subtracted from the checkbook balance.

Find the amount of the charge card fee for each month if the fee is:

Percent Charged	For Sales of
6%	$100 to $1,999
4%	$2,000 to $4,000
1%	over $4,000

	Month	Charge Card Sales	Amount of Fee
1.	January	$709.53	________
2.	February	$2,189.00	________
3.	March	$564.00	________

EVALUATING YOUR SKILLS

The following problems evaluate your skills in the material covered in Units 1 through 10. Some of the problems will evaluate your ability to work problems mentally while others will evaluate your ability to use your calculator. Work the problems as quickly as you can.

1. Which number is smaller, 1.059, 1.095, or 1.0089? __________
2. Find the lowest common denominator of the following fractions: $\frac{5}{8}$, $\frac{5}{12}$, $\frac{7}{20}$ __________
3. Change $\frac{153}{30}$ to a mixed number. __________
4. Change $\frac{1}{5}$% to a decimal. __________
5. Which is greater, $\frac{7}{15}$ or $\frac{12}{25}$? __________
6. 84,664 + 96,980 − 47,397 + 85,866 − 20,100 = __________
7. (237 × 15.7) − 244.8 = __________
8. What is $15\frac{1}{2}$% of $1,220? __________
9. What percent of 60 is 72? __________
10. 12% of what number is 6? __________
11. An average of five numbers is 120. Given numbers are 127, 125, 110, and 115. Find the missing number. __________
12. Mentally multiply .9 × 1,000. __________
13. Mentally multiply 6 × .01. __________
14. An item costs $1.47 for 7 oz. Find the price per lb. __________
15. Strawberries are sold in a two liter box. How many quarts is this? __________

APPLYING YOUR KNOWLEDGE

Solve the following problems using the information given.

1. Your bank statement and checkbook balances are not in agreement. You discovered an error in recording a check written for $232.79. You recorded $223.97 in your check register.
   a. If there are no other errors, by how much will the adjusted bank balance differ from the adjusted checkbook balance? ______________
   b. Should you add or subtract the difference in your check register? ______________

2. An artist who specializes in bronze statues of Western art sells his works overseas. His total foreign bank card sales for December were $5,890.29. If the bank charges a fee of $5\frac{1}{4}$% for the service, what amount will be deducted from the artist's checking account? ______________

3. Compare the following check register with the partial bank statement. Put a check mark beside the checks and deposits processed by the bank. Prepare a bank reconciliation statement. Any adjustments to the check register should be recorded on the blank lines in the register.

BANK STATEMENT

Previous Balance	No.	Check Amount	Deposit Amount	Service Charge	New Balance
$3,139.21	69	122.84			3,016.37
	70	34.95			2,981.42
	71	240.00			2,741.42
	73	216.90	650.00		3,174.52
	75	89.60		6.00	3,078.92

PLEASE BE SURE TO **DEDUCT** ANY PER ITEM CHARGES OR SERVICE CHARGES THAT MAY APPLY TO YOUR ACCOUNT

ITEM NO.	DATE	PAYMENT ISSUED TO OR DESCRIPTION OF DEPOSIT	AMOUNT OF PAYMENT		√	AMOUNT OF DEPOSIT		BALANCE 3,139	21
67	11/3	C-W Gas Company	83	21				3,056	00
68	11/4	Joe Elmers	16	00				3,040	00
69	11/7	Auto & Tire Supply	122	84				2,917	16
70	11/10	Munchies, Inc.	43	95				2,873	21
71	11/14	Rent	240	00				2,633	21
72	11/15	S-W Telephone Co.	32	18		650	00	3,251	03
73	11/17	Car Insurance Brokers	216	90				3,034	13
74	11/23	Jean's Gift Shop	27	67				3,006	46
75	11/25	Audrey's T.V. Repair Shop	89	60				2,916	86

BANK RECONCILIATION
December 5, 19--

4. Arthur Boxwell owns a flower shop. He received his bank statement showing a balance of $3,424.88. His checkbook balance is $3,802.39. There is a charge card fee of $32.80 on the statement. Two outstanding deposits are $150.65 and $278.42. There is no service charge. He earned $14.58 in interest for the month. Outstanding checks are #256 for $27.60 and #259 for $42.18. Prepare a bank reconciliation for Arthur on January 1.

BANK RECONCILIATION
January 1, 19--

5. George Van Garden's checkbook balance is $1,251.63. His bank statement for the same day is $1,350.80. Along with the bank statement was a notice that an electronic payment had been made to his insurance company in the amount of $85.72. There was a service charge of $7.50 and interest earned in the amount of $5.04. Outstanding checks are #410 for $24.94 and #415 for $32.41. A deposit of $130 was not recorded in his check register. Prepare a bank reconciliation on June 1.

BANK RECONCILIATION
June 1, 19--

ADVANCED APPLICATION

Assume you are the accountant for the Amber Company. Verify the accuracy of the Amber Company's October register, which follows. Place a check mark beside any incorrect balance. Prepare a bank reconciliation for October 31 using this information: bank statement balance is $52,947.66; automatic (EFT) payment for $782.00; service charge of $6.78; bank card fee of $125.70; customer's check for $167.19 returned insufficient; outstanding deposit of $4,762.18; outstanding checks are #1565 for $72.00, #1580 for $729.48, #1582 for $638.95, and #1586 for $242.89; and Check #1575 was written for $674.30 but recorded in check register as $647.30.

BANK RECONCILIATION
October 31, 19--

PLEASE BE SURE TO **DEDUCT** ANY PER ITEM CHARGES OR SERVICE CHARGES THAT MAY APPLY TO YOUR ACCOUNT

ITEM NO.	DATE	PAYMENT ISSUED TO OR DESCRIPTION OF DEPOSIT	AMOUNT OF PAYMENT		√	AMOUNT OF DEPOSIT		BALANCE	
								42,147	85
	10/1					1,722	33	43,870	18
1562	10/1	Rent	700	00				43,170	18
1563	10/1	Insurance	79	64				43,090	54
1564	10/1	Salaries	1,793	87				41,296	67
1565	10/3	Yellow Page Ad	72	00		1,600	60	42,891	27
1566	10/6	Window Service	12	00				42,879	27
1567	10/8	Freight	19	92				42,859	35
1568	10/9	Utilities	20	62				42,838	73
1569	10/9	Truck expense	75	22		2,200	00	44,963	51
1570	10/9	Telephone	140	81				44,822	70
1571	10/9	Miscellaneous Supplies	371	93				44,450	77
1572	10/9	Wallpaper	262	47				44,188	30
1573	10/9	Paint	203	57				43,974	73
1574	10/9	Flowers	15	60				43,959	13
1575	10/9	Paint	647	30				43,311	83
1576	10/9	Wallpaper	283	47				43,028	36
1577	10/10	Frames	95	41		3,015	85	45,948	80
1578	10/11	Wallpaper	420	96				45,527	84
1579	10/15	Salaries	1,793	87				43,733	97
1580	10/15	Federal Tax deposit	729	48				43,004	49
1581	10/15	Utilities	83	59		4,604	66	47,525	56
1582	10/17	Wallpaper	638	95				46,886	61
1583	10/19	Truck expense	121	50		3,982	11	50,757	22
1584	10/19	Sales Tax deposit	373	34				50,383	88
1585	10/22	Postage	20	50		4,112	39	54,475	77
1586	10/29	Ad Valorem Tax	242	89				54,232	88
1587	10/29	Salaries	1,793	87				52,439	01
	10/31					4,762	18	57,201	19

Wages and Salaries

"You can if you will. Don't be a quitter."

Payroll is an important part of every business' records. Although there are many ways to pay employees—salary, wages, piecework, commission—each employee's gross earnings, deductions, and net earnings must be calculated.

After studying this unit, you should be able to:

1. compute regular time and overtime gross earnings
2. determine total hours worked using time cards
3. compute gross earnings for piece-rate plan

MATH TERMS AND CONCEPTS

The following terminology is used in this unit. Become familiar with the meaning of each term so that you understand its usage as you develop math skills.

1. **Double Time**—two times the regular hourly rate.
2. **Gross Earnings**—amount earned before any deductions are made.
3. **Overtime**—hours worked beyond regular hours.
4. **Salary**—a fixed amount of pay for a certain pay period, usually for a week, month, or year.
5. **Time Card**—a record of the number of hours worked.
6. **Time Clock**—a clock that records an employee's arrival and departure times on a time card when the time card is inserted into the clock.
7. **Time and a Half**—one and a half times the regular hourly rate.
8. **Wages**—an amount of pay based on actual hours worked or actual production.

Hourly Wages

Many businesses require employees to punch in on a **time clock**. The time clock accurately records when an employee arrives and leaves. Late arrival or early departure is indicated. At the end of a time period, usually one or two weeks, the total hours worked are calculated on each time card. **Gross earnings** are computed by multiplying the number of hours worked by the **wages** per hour.

Example: Sara Henley works for Martinez Manufacturing Co. She earns \$4.50 per hour for a 40-hour workweek. If she works a regular workweek, her gross earnings are calculated as: \$4.50 × 40 = \$180.

Determine the gross earnings for each of the following employees.

	Name	Total Hours	Hourly Wage	Gross Earnings
1.	Wong	40	\$6.70	________
2.	Stayton	35.8	\$7.25	________
3.	Whitehead	39	\$7.50	________
4.	Legrande	34.1	\$9.50	________
5.	Suggs	38.5	\$8.75	________

Overtime

Most nonmanagerial or nonprofessional employees are paid extra wages for any time worked over their regular workweek. **Overtime** is figured on a daily or a weekly basis. On a daily basis, time worked over eight hours a day is overtime. On a weekly basis, time worked over 40 hours is overtime.

Example: An employee's regular workweek is 40 hours. The total hours worked are 47. Gross earnings are figured on 7 *overtime* hours plus the regular 40 hours.

Most companies pay **time and a half** for overtime. The regular hourly wage rate is divided by 2 and this amount is added to the hourly wage to find the overtime rate.

Example: Regular hourly wage is \$4. Time and a half is \$6.

\$4 ÷ 2 = \$2 (half of regular earnings)
\$4 + \$2 = \$6 (time and a half)

The same overtime rate can be figured quickly by multiplying the regular rate by 1.5.

\$4 × 1.5 = \$6

Example: James Mecker worked 47 hours at \$4.20 an hour with time and a half for all hours worked over 40. His gross earnings are calculated as follows:

40 hours regular time
7 hours overtime
\$4.20 × 40 = \$168 (regular pay)
\$4.20 × 1.5 × 7 = \$44.10 (overtime pay)
\$168 + \$44.10 = \$212.10 (gross pay)

Use multifactor multiplication to calculate overtime pay plus accumulative multiplication to obtain gross earnings.

Compute gross earnings for the following individuals. All hours worked over 40 are paid at time and a half.

	Employee	Total Hours	Hourly Rate	Regular Earnings	Overtime Earnings	Gross Earnings
1.	Abney	44	$8.20	________	________	________
2.	Brotherton	43	$8.20	________	________	________
3.	Devillez	37	$7.25	________	________	________
4.	Garcia	44	$7.25	________	________	________
5.	Koonce	45	$7.25	________	________	________

Time Cards

A **time card** is used to record the time an employee arrives at work and leaves work. The time clock is an accurate way to record the employee's arrival at the beginning of the work day, the time taken for lunch, and the departure time.

The time clock usually records partial hours worked to two decimal places. Each company establishes its own guidelines for determining when an employee begins receiving overtime pay and its own policies regarding early and late arrival.

Study the time card for Brenda Perkins in Illustration 12-1. She is paid for the exact number of hours and minutes she works. She receives overtime pay after working 40 hours during a week. The number of hours and minutes between each IN and OUT segment are added to arrive at the total.

* ***Remember, there are 60 minutes in an hour.***

DAY	IN	OUT	IN	OUT	TOTAL
MONDAY	7:59	12:02	1:00	5:03	8:06
TUESDAY	8:01	12:00	1:00	5:00	7:59
WEDNESDAY	8:06	11:55	12:49	4:45	7:45
THURSDAY	7:58	12:00	1:00	4:00	7:02
FRIDAY	8:29	12:01	1:00	4:45	7:17
				TOTAL	38:09

Illustration 12-1. Time Card.

Some companies calculate time worked based on portions of an hour. The hour may be divided into four equal parts with each fifteen minutes equaling one-fourth of an hour. An employee is penalized for each one-fourth of an hour not worked. Another method is to divide the hours into ten equal parts with one-tenth of an hour equaling six minutes. The employee is penalized for each one-tenth of an hour not worked. The employee is paid overtime based on the tenths of an hour worked over eight hours in any day.

Remember, each business establishes its own policies regarding early and late arrival.

Complete the following time cards for Littleton Company. Policies on computing time are as follows.

1. Regular workday is 8:00 to 12:00 and 1:00 to 5:00.
2. Hours worked over 40 a week are paid at time and a half.
3. Employees are paid for the exact number of hours and minutes worked.

1.

DAY	IN	OUT	IN	OUT	HOURS WORKED
M	7:56	12:00	1:00	5:00	________
TU	8:35	12:00	1:00	5:00	________
W	7:59	11:46			________
TH	8:00	12:00	1:00	6:30	________
F	8:00	11:58	1:19	5:00	________

Total Hours Worked ________

Total Overtime Hours Worked ________

2.

DAY	IN	OUT	IN	OUT	HOURS WORKED
M	8:00	12:00	12:55	5:00	________
TU	8:03	12:00	1:01	4:58	________
W	8:16	12:01	1:00	5:00	________
TH	8:00	11:45	1:00	5:00	________
F	7:58	11:58	1:00	6:30	________

Total Hours Worked ________

Total Overtime Hours Worked ________

3.

DAY	IN	OUT	IN	OUT	HOURS WORKED
M	7:30	12:01	1:02	8:20	______
TU	7:59	12:00	1:00	5:00	______
W	8:00	12:00	1:00	8:15	______
TH	7:30	12:00	12:59	6:45	______
F	8:00	12:00	1:00	5:00	______

Total Hours Worked ______

Total Overtime Hours Worked ______

4.

DAY	IN	OUT	IN	OUT	HOURS WORKED
M	8:00	12:00	1:00	5:04	______
TU	7:59	12:00	12:57	5:00	______
W	7:07	11:59	1:00	5:00	______
TH	8:00	12:01	1:02	5:00	______
F	7:58	12:02	1:04	5:30	______

Total Hours Worked ______

Total Overtime Hours Worked ______

Piece-rate Wages

Some employees are paid for every item of merchandise they produce instead of an hourly wage. Labor Department regulations require that the piece-rate wage be equal to, or more than, minimum wage.

To determine gross earnings, multiply the rate per item by the number produced.

Example: Tony Rose is paid \$.60 for every toy he assembles. He assembled 82 on Monday, 80 on Tuesday, 79 on Wednesday, 83 on Thursday, and 81 on Friday. What is his gross pay?

$82 + 80 + 79 + 83 + 81 = 405$ (total units produced)
$405 \times \$.60 = \$243$ (gross pay)

Determine the total units produced and gross earnings for each employee if the piece-rate wage is \$.55 for each unit assembled.

	Employee	Mon	Tues	Wed	Thurs	Fri	Total	Gross Earnings
1.	Storme, W.	77	79	60	82	80	____	____________
2.	Roseler, K.	93	91	88	90	87	____	____________
3.	Tanner, R.	86	82	89	87	86	____	____________
4.	Tran, Q.	93	90	92	98	95	____	____________
5.	Whit, R.	81	72	77	80	76	____	____________

As an incentive for employees to produce more, an employer may offer an employee a higher rate for units produced over a minimum number.

Example: Picken Tool Co. pays \$.45 for the first 400 units produced a week, \$.47 for 401-450 units produced, and \$.50 for all units produced over 450. If an employee produced 475 units, the gross earnings would be \$216. The gross earnings are calculated this way:

$400 \times \$.45 = \$180$
$50 \times \$.47 = \$23.50$
$25 \times \$.50 = \$12.50$
$\$180 + \$23.50 + \$12.50 = \$216$

Determine gross earnings for the following Picken Tool Co. employees. Use the piece-rate scale given in the preceding example.

	Employee	M	Tu	W	Th	F	Total Units	Earnings at .45	Earnings at .47	Earnings at .50	Gross Earnings
6.	Calder	80	83	81	85	90	________	________	________	________	________

	Employee	M	Tu	W	Th	F	Total Units	Earnings at .45	.47	.50	Gross Earnings
7.	Neal	93	92	96	98	93	______	______	______	______	______
8.	Rice	79	75	80	77	76	______	______	______	______	______
9.	Rouls	86	83	92	91	97	______	______	______	______	______
10.	Sewell	87	95	97	98	99	______	______	______	______	______

Salaries

A **salary** is a fixed amount paid for a certain period of time. The pay period may be weekly, biweekly, semimonthly, or monthly. The annual salary is the total amount of earnings for a year. The salary per pay period is found by dividing the annual salary by the number of pay periods per year.

Example: Alan Walker is paid an annual salary of $18,840. He has a choice of receiving his paycheck weekly, biweekly, semimonthly, or monthly. His salary for each pay period would be calculated as follows:

Weekly:	$18,840 ÷ 52 = $362.31
Biweekly:	$18,840 ÷ 26 = $724.62
Semimonthly:	$18,840 ÷ 24 = $785.00
Monthly:	$18,840 ÷ 12 = $1,570.00

Find each of the following employee's salary for the pay period indicated.

	Employee	Annual	Pay Period	Pay Periods	Salary
1.	Alspah, P.	$10,200	Semimonthly	______	______
2.	Bostick, A.	$11,400	Monthly	______	______
3.	Felton, Z.	$8,450	Biweekly	______	______

	Employee	Annual	Pay Period	Pay Periods	Salary
4.	Garris, R.	$11,700	Weekly	________	________
5.	Rozen, W.	$15,600	Monthly	________	________

EVALUATING YOUR SKILLS

The following problems evaluate your skills in the material covered in Units 1 through 11. Some of the problems will evaluate your ability to work problems mentally while others will evaluate your ability to use your calculator. Work the problems as quickly as you can.

1. Round 321.9972 to the nearest tenth. ________

2. Change $3\frac{2}{3}$ to an improper fraction. ________

3. Add $\frac{1}{2}$ and $\frac{1}{3}$. ________

4. Multiply $\frac{1}{5}$ by $\frac{2}{7}$. ________

5. What % of 200 is 300? ________

6. Express $\frac{9}{11}$ as a decimal to the nearest hundredth. ________

7. Multiply 10,000 by $.10. ________

8. Divide 2.76 by 10. ________

9. Estimate the quotient: 759,432 ÷ 37. ________

10. Express $\frac{1}{50}$ as a percent. ________

11. Divide .9261 by .21. ________

12. How many liters equal 137 gallons? ________

13. Shirley's Dress Shop sold 30 dresses at $45.97 each, 12 at $32.20 each, and 5 at $24.60 each. What was the average price per dress? ______

14. Burt Rumsey had a checkbook balance of $615.11 on September 1. His bank statement balance was $516.20. Outstanding checks were $102.14 and $34.97. An EFT of $68.02 for an insurance payment was deducted from the bank statement. There was an outstanding deposit of $168. What is the reconciled balance? ______

15. An invoice dated July 17 was paid on August 1. The amount of the invoice was $342. The terms were 3/15, n/60. What was the amount of the check? ______

APPLYING YOUR KNOWLEDGE

Solve the following problems. Read each one carefully.

1. Becky Hoggard earned $48 in overtime earnings last week. This was $\frac{1}{6}$ of her total earnings. What was her total gross pay? ______

2. Designer's Showcase pays $7.30 an hour for an 8-hour day with time and a half for over 8 hours a day. During one week Mark Lavitt worked the following hours: Monday, 8; Tuesday, 9.5; Wednesday, 6; Thursday, 10; Friday, 7. What was Mark's total gross pay? ______

3. Ackerman is paid $.33 for each piece assembled. His record shows 142 pieces completed on Monday, 148 completed on Tuesday, 151 on Wednesday, 144 on Thursday, and 140 on Friday. Twenty-one pieces were rejected. (He is not paid for any pieces rejected.) What was Ackerman's gross earnings for the week? ______

4. Kevin works a 40-hour week at $10.50 an hour with time and a half for overtime. Last week he worked $47\frac{3}{4}$ hours. Find Kevin's gross pay.

5. Joan Newsom has two job offers. Job A pays $535 semimonthly. Job B pays $255 weekly.
   a. Which job pays the most salary?

   b. How much more per year does the better offer pay?

6. Jerry Vin's time card showed that he worked from 7:51 to 12:00 and from 1:00 to 4:55. He is scheduled to work from 8:00 to 12:00 and from 1:00 to 5:00. How many hours did he work?

7. Stern is offered a job at $220 a week. Another job offer is for $935 a month.
   a. Which salary is the better offer?

   b. By how much?

8. Steve Johnson is paid $.75 for each unit he produces. During one week he produced 525 units. Determine his hourly rate, based on a 40-hour week.

ADVANCED APPLICATIONS

Using the information given, solve the following problems.

1. Lucia Lopez earned $1,260 during the month of November. This was $\frac{1}{5}$ more than what she earned during the month of October. How much did she earn in October?

2. Bill Henry earns $12.82 per hour plus $.25 per hour travel allowance. On March 5 the travel allowance was increased to $.36 per hour. All hours he works over 8 hours per day are paid at one and a half. How much did he earn for the week if he worked the following hours?

Day	Hours Worked
March 3	15.4
March 4	9.9
March 5	12.0
March 6	8.0
March 7	10.3

____

3. A union contract specifies employees will be paid $12.50 per hour. Any hours worked from 7 p.m. to 7 a.m. receive an additional $1.40 per hour. Any hours worked over 8 hours per day are paid at time and a half. Time worked on Saturday or Sunday is double time. The employees are given 1 hour for lunch without pay during an 8-hour work period. Find the gross pay for the following employees.

a. John Higgins ______________

Day	IN	OUT	Regular Hours	Night Hours	Time-and-a-half Overtime	Double Overtime
Monday	7:00 a.m.	5:00 p.m.				
Tuesday	10:00 p.m.	7:00 a.m.				
Wednesday	7:30 a.m.	6:00 p.m.				
Thursday	day	off				
Friday	7:00 a.m.	6:00 p.m.				
Saturday	10:00 a.m.	7:00 p.m.				
Sunday	day	off				
Total Hours						

b. Judy Fulkers ______________

Day	IN	OUT	Regular Hours	Night Hours	Time-and-a-half Overtime	Double Overtime
Monday	8:00 a.m.	5:00 p.m.				
Tuesday	1:00 p.m.	10:00 p.m.				
	Lunch from	5-6 p.m.				
Wednesday	4:00 p.m.	1:00 a.m.				
	Lunch from	8-9 p.m.				
Thursday	day	off				
Friday	day	off				
Saturday	10:00 a.m.	7:00 p.m.				
Sunday	8:00 a.m.	5:00 p.m.				
Total Hours						

c. Ruth Zackis ______________

Day	IN	OUT	Regular Hours	Night Hours	Time-and-a-half Overtime	Double Overtime
Monday	8:00 a.m.	5:00 p.m.				
Tuesday	7:00 a.m.	4:00 p.m.				
Wednesday	7:00 p.m.	4:00 a.m.				
	Lunch from	11-12				
Thursday	7:30 a.m.	5:30 p.m.				
Friday	7:00 a.m.	7:00 p.m.				
Saturday	day	off				
Sunday	day	off				
Total Hours						

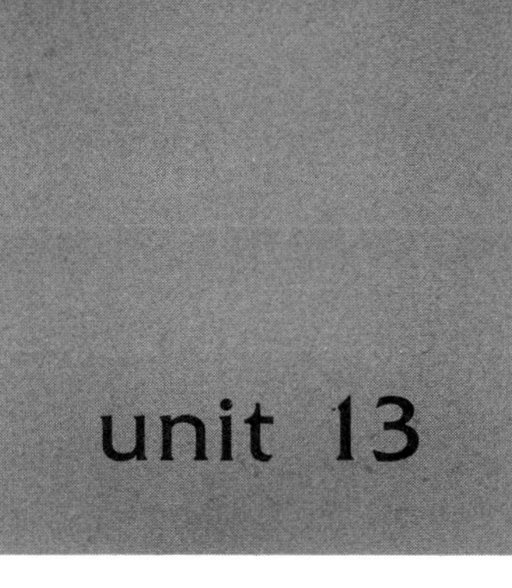

Commissions and Deductions

"The man who watches the clock usually remains one of the hands."

Commissions or commissions plus salary are often paid as an incentive to increase employee production. Accurate payroll records must be kept. Forms must be completed to be sent to state and federal tax offices.

After studying this unit, you should be able to:

1. compute gross earnings on straight commission, commission plus salary, and graduated commission
2. compute net earnings
3. complete payroll reports

MATH TERMS AND CONCEPTS

The following terminology is used in this unit. Become familiar with the meaning of each term so that you understand its usage as you develop math skills.

1. **Commission**—a percentage of the total amount of sales.
2. **Dependent**—an individual who can be claimed as an exemption for tax purposes.
3. **Exemptions**—allowances for dependents and for oneself as a deduction from gross income.
4. **Exemption from Withholding**—an employee who earns an amount less than a predetermined amount set by the IRS and who can have no income tax withheld.
5. **FICA Tax**—Federal Insurance Contributions Act tax.
6. **Payroll Deductions**—amounts subtracted from gross earnings to obtain net earnings.
7. **Quarter**—a three-month period.

Commissions

Many salespersons are paid according to how much is sold. A percent of total sales is the salesperson's **commission**, or gross earnings.

Straight Commission

A salesperson who receives only a commission on what is sold with no salary is paid a *straight commission.*

Can you see that total sales is the base and you are finding the percentage? (B × R = P)

Example: Sue Zulokis earns a 7% straight commission on her total sales. She sold $6,200 in one week; her gross earnings were $434.

$6,200 × 7% = $434

Find the commission earned on each of the following sales.

	Amount	Rate of Commission	Gross Earnings
1.	$900	5%	______
2.	$2,235	9%	______
3.	$6,281	$4\frac{1}{2}$%	______
4.	$5,790	12%	______
5.	$800	$6\frac{1}{2}$%	______

Commission Plus Salary

A salesperson may receive a salary *plus* a commission on sales.

Example: Jacob Miller receives $95 a week plus a 4% commission on all sales. He sells $920 in a week; his gross earnings are $131.80.

$920 × 4% = $36.80
$36.80 + $95 = $131.80

Find the gross earnings on each of the following sales.

	Commission Earned	Salary Earned	Total Sales	Gross Earnings
1.	3% on all sales	$120 a week	$3,220 for a week	______
2.	5% on all sales	$450 a month	$12,872 for a month	______
3.	3% on all sales over $1,000	$170 a week	$2,275 for a week	______
4.	4.5% on all sales over $3,500	$625 a month	$8,326 for a month	______
5.	5% on all sales over $2,200	$500 a month	$7,320 for a month	______

Graduated Commission

Commissions are often paid on a graduated scale. As sales increase to a given level the rate will also increase. This is to "reward" employees who achieve certain sales levels.

Example: Arnold Company pays its employees a base salary of $600 a month plus a commission on a graduated scale as follows:

2% on first $10,000 sales
3% on all sales over $10,000 through $14,000
4% on all sales over $14,000

Example: Mrs. Klentos, a salesperson for Arnold Company, sells $16,300 during February. Her earnings are calculated using the graduated scale:

$10,000 × 2% = $200
$4,000 × 3% = $120
$16,300 − $14,000 = $2,300
$2,300 × 4% = $92
$200 + $120 + $92 = $412 (total commission)
$600 + $412 = $1,012 (gross earnings)

Use accumulative multiplication.

Find the gross earnings for the following salespersons. In addition to a weekly salary, a 2% commission is earned on all sales up to $3,000; $2\frac{1}{2}$% on sales over $3,000 through $5,000; and 3% on all sales over $5,000.

	Salesperson	Weekly Sales	Weekly Salary	Commission at 2%	Commission at $2\frac{1}{2}$%	Commission at 3%	Gross Earnings
1.	Bush, B.	$3,759	$250	________	________	________	________
2.	O'Neill, J.	$5,500	$250	________	________	________	________
3.	Shuberg, R.	$2,700	$300	________	________	________	________
4.	Tse, C.	$4,800	$300	________	________	________	________
5.	Volz, T.	$4,320	$300	________	________	________	________

Employers are required by law to withhold from employees' pay amounts to cover federal income tax, social security tax, and for some states, state income tax. Other **payroll deductions** may include insurance, union dues, and savings bonds.

Federal Withholding Tax

The amount of federal income tax withheld from a paycheck is determined by three factors.

1. *Marital Status*—single (not married) or married.
2. *Number of exemptions*—One **exemption** is allowed for the taxpayer, the taxpayer's spouse if not claimed by the spouse, and each **dependent**. A dependent may be claimed by only one taxpayer. The exception is a student who may be claimed by both the parent and the student.
3. *Total amount earned*—an employee's gross earnings. Many students working part-time jobs will not earn enough to have a tax liability. They may claim an **exemption from withholding** with their employer. The employer will not withhold any income tax.

Tax tables are used to find the amount to be withheld from an employee's gross earnings. Tables are available for daily, weekly, biweekly, semimonthly, and monthly time periods. The tax tables in Illustration 13-1 show the deductions to be made for various weekly salaries. The top table is used for a married person, the bottom table is used for a single person. Please keep these tables; they will be used in this unit and in later units.

Example: George Baker earns $275 a week. He is single and claims only himself as an exemption. Read down the column for the wages that read at least $270 but less than $280. Read across to the column labeled 1. George will have $32 withheld from his gross earnings for federal income tax.

Use the tables in Illustration 13-1, page 227, to find the amount to be withheld from the weekly salary of each of the following people.

	Employee	Marital Status	Exemptions	Salary	Tax
1.	Burns, B.	S	1	$410.00	__________
2.	Dobbs, J.	M	3	$462.50	__________
3.	Ennis, C.	M	2	$513.18	__________
4.	Gober, J.	S	0	$352.20	__________
5.	Jared, C.	M	4	$514.40	__________

Social Security Taxes

The Federal Insurance Contributions Act (FICA) provides pensions after retirement, disability payments, survivors' benefits, and medical and hospital benefits. The **FICA tax** or *social security tax* is a percentage of a maximum base. Once an employee's earnings pass the maximum earnings, FICA tax deductions stop for the rest of the year. The percent and base, set by Congress, have steadily increased. Payroll personnel must keep accurate records of each employee's earnings.

MARRIED Persons—WEEKLY Payroll Period

And the wages are-		And the number of allowances claimed is-										
At least	But less than	0	1	2	3	4	5	6	7	8	9	10
		The amount of income tax to be withheld shall be-										
270	280	32	26	20	15	9	3	0	0	0	0	0
280	290	34	28	22	16	10	5	0	0	0	0	0
290	300	35	29	23	18	12	6	0	0	0	0	0
300	310	37	31	25	19	13	8	2	0	0	0	0
310	320	38	32	26	21	15	9	3	0	0	0	0
320	330	40	34	28	22	16	11	5	0	0	0	0
330	340	41	35	29	24	18	12	6	1	0	0	0
340	350	43	37	31	25	19	14	8	2	0	0	0
350	360	44	38	32	27	21	15	9	4	0	0	0
360	370	46	40	34	28	22	17	11	5	0	0	0
370	380	47	41	35	30	24	18	12	7	1	0	0
380	390	49	43	37	31	25	20	14	8	2	0	0
390	400	50	44	38	33	27	21	15	10	4	0	0
400	410	52	46	40	34	28	23	17	11	5	0	0
410	420	53	47	41	36	30	24	18	13	7	1	0
420	430	55	49	43	37	31	26	20	14	8	3	0
430	440	56	50	44	39	33	27	21	16	10	4	0
440	450	58	52	46	40	34	29	23	17	11	6	0
450	460	59	53	47	42	36	30	24	19	13	7	1
460	470	61	55	49	43	37	32	26	20	14	9	3
470	480	62	56	50	45	39	33	27	22	16	10	4
480	490	64	58	52	46	40	35	29	23	17	12	6
490	500	65	59	53	48	42	36	30	25	19	13	7
500	510	67	61	55	49	43	38	32	26	20	15	9
510	520	68	62	56	51	45	39	33	28	22	16	10
520	530	70	64	58	52	46	41	35	29	23	18	12
530	540	71	65	59	54	48	42	36	31	25	19	13
540	550	73	67	61	55	49	44	38	32	26	21	15
550	560	74	68	62	57	51	45	39	34	28	22	16
560	570	76	70	64	58	52	47	41	35	29	24	18

SINGLE Persons—WEEKLY Payroll Period

And the wages are-		And the number of allowances claimed is-										
At least	But less than	0	1	2	3	4	5	6	7	8	9	10
		The amount of income tax to be withheld shall be-										
240	250	34	28	22	16	11	5	0	0	0	0	0
250	260	35	29	24	18	12	6	0	0	0	0	0
260	270	37	31	25	19	14	8	2	0	0	0	0
270	280	38	32	27	21	15	9	3	0	0	0	0
280	290	40	34	28	22	17	11	5	0	0	0	0
290	300	41	35	30	24	18	12	6	1	0	0	0
300	310	43	37	31	25	20	14	8	2	0	0	0
310	320	44	38	33	27	21	15	9	4	0	0	0
320	330	46	40	34	28	23	17	11	5	0	0	0
330	340	47	41	36	30	24	18	12	7	1	0	0
340	350	49	43	37	31	26	20	14	8	2	0	0
350	360	50	44	39	33	27	21	15	10	4	0	0
360	370	52	46	40	34	29	23	17	11	5	0	0
370	380	53	47	42	36	30	24	18	13	7	1	0
380	390	56	49	43	37	32	26	20	14	8	3	0
390	400	58	50	45	39	33	27	21	16	10	4	0
400	410	61	52	46	40	35	29	23	17	11	6	0
410	420	64	53	48	42	36	30	24	19	13	7	1
420	430	67	56	49	43	38	32	26	20	14	9	3
430	440	70	59	51	45	39	33	27	22	16	10	4
440	450	72	62	52	46	41	35	29	23	17	12	6
450	460	75	64	54	48	42	36	30	25	19	13	7
460	470	78	67	56	49	44	38	32	26	20	15	9
470	480	81	70	59	51	45	39	33	28	22	16	10
480	490	84	73	62	52	47	41	35	29	23	18	12
490	500	86	76	65	54	48	42	36	31	25	19	13
500	510	89	78	68	57	50	44	38	32	26	21	15
510	520	92	81	70	60	51	45	39	34	28	22	16
520	530	95	84	73	62	53	47	41	35	29	24	18
530	540	98	87	76	65	54	48	42	37	31	25	19

Illustration 13-1. Tax Tables.

Social security tax is figured by multiplying the gross earnings by the tax rate. A tax rate of 7.65% and a maximum base of $51,300 gross earnings will be used in this text.

Example: Ruth Posey earned $345 for the week ending July 3. She has not earned more than $51,300. Therefore, all her earnings are subject to FICA tax. Her deduction for social security tax will be $26.39.

$345 × 7.65% = $26.39

7.65% can be a constant using .0765.

Find the social security tax for each of the following. Use 7.65% as the tax rate.

	Gross Earnings	Total Earned to Date	Tax
1.	$2,500 a month	$15,000.00	______
2.	$576 a week	$16,153.76	______
3.	$2,916.67 a month	$29,166.70	______
4.	$1,833.33 a month	$20,166.63	______
5.	$4,775 a month	$52,525.00	______

Federal Tax Deposit

Each month a tax deposit is made to the employer's bank. The deposit includes the federal income tax and social security tax deducted from the gross earnings of all employees. At the end of each **quarter**, a Form 941 is filed with the Internal Revenue Service (IRS) showing the amount of each month's tax deposit. Each percentage deducted from gross earnings for social security is matched by the employer. The employer's contribution is also 7.65%. The total amount deposited for both the employer's portion and the employee is 15.3%.

Example: The Zenix Company's total gross earnings for all employees for the month were $20,249.52. No employee has earned the maximum social security salary. The social security tax deposit for the month is $3,098.18.

$20,249.52 × 15.3% = $3,098.18

The deposit, called the *federal tax deposit*, is $3,098.18 plus any federal income tax deducted from employees' gross earnings.

Remember: The employer must contribute to Social Security an amount equal to 7.65% of each employee's gross earnings.

Determine the federal tax deposit in January for each employer.

Employer	Employees' Gross Earnings	Federal Income Tax	Social Security Tax	Federal Tax Deposit
1	$15,782	$631.28	______	______

Employer	Employees' Gross Earnings	Federal Income Tax	Social Security Tax	Federal Tax Deposit
2	$22,980	$919.20	________	________
3	$57,240	$2,289.60	________	________
4	$17,243	$689.72	________	________
5	$29,617	$1,184.68	________	________

Payroll Register

A payroll register form is used to show the earnings and deduction information for all employees of a company. When the payroll register is completed, it will look similar to Illustration 13-2.

			Gross Earnings					Deductions				
Name	Marital Status	Exemp-tions	Hours	Hourly Wages	Regular Wage	OT Wage	Total	FICA Tax	Federal Tax	State Tax	Insurance	Total Net Pay
Ashley, T. C.	S	1	40	$8.50	$340.00	$ -0-	$340.00	$26.01	$43.00	$18.00	$ -0-	$252.99
Buoy, C. W.	S	2	43	8.75	350.00	39.38	389.38	29.79	43.00	18.00	-0-	298.59
Butler, F. E.	M	4	40	8.75	350.00	-0-	350.00	26.78	21.00	8.00	20.30	273.92

Illustration 13-2. Payroll Register.

Computers are handling the payroll procedures for a growing number of businesses. Marital status, number of exemptions, hours worked, hourly wage or salary, and the tax rate or tables are programmed into the computer. The computer will then calculate gross earnings, deductions, and net earnings.

Complete the following simplified payroll register. Figure FICA tax by multiplying gross earnings by 7.65%. Then calculate net earnings.

	Name	Gross Earnings	Income Tax	FICA Tax	Insurance	Net Earnings
1.	Baker, E.	$760	$177	$58.14	$87	________
2.	Carroll, F.	$810	$159	$61.97	$87	________
3.	Fong, G.	$560	$84	$42.84	$87	________
4.	Smith, S.	$820	$169	$62.73	$87	________
5.	Taret, L.	$730	$173	$55.85	$87	________

EVALUATING YOUR SKILLS

The following problems evaluate your skills in the material covered in Units 1 through 12. Some of the problems will evaluate your ability to work problems mentally while others will evaluate your ability to use your calculator. Work the problems as quickly as you can.

1. Divide to three decimal places: 17,500 ÷ 17,000 __________
2. Add $6\frac{1}{4} + 3\frac{1}{3} + 2\frac{1}{6}$. __________
3. .463 × 1,000 = __________
4. Show $1\frac{1}{4}\%$ as a decimal. __________
5. 24 at $3.14 =
   16 at $4.04 =
   23 at $13.29 =
   Total __________
6. Change 1.3% to a fraction. __________
7. Write .479 as a percent. __________
8. What is the price per pound for instant coffee that sells for $4.20 for a 6 oz jar? __________
9. A salary of $176 a week is equal to how much a month? __________
10. Convert 5 quarts to liters. __________
11. 240 is what percent of 9,600? __________
12. John made an 85, 87, 83, and 78 on the first four tests. What must he make on the fifth test if he wants to average 85 on all five tests? __________
13. Estimate the product of 4,892.10 × 52. __________
14. On June 30, Jeff Stone had a checkbook balance of $2,240.24. His bank statement on that date showed a balance of $2,343.18. Outstanding checks were for $22.19, $35.29, and $117.80. Jeff discovered a cancelled check for $67.84 he had not recorded in the check register. A service charge for $4.50 had been deducted on his bank statement. Prepare a reconciliation statement.

BANK RECONCILIATION
June 30, 19--

APPLYING YOUR KNOWLEDGE

Solve the following problems. Read each one carefully.

1. Jean is married and claims four exemptions.
   a. Use the tax table on page 227 to find federal income tax if her gross pay is $542.06. ______
   b. Compute FICA tax. ______
   c. Jean has a $2 bond deducted from each paycheck. Find Jean's net pay. ______

2. A salesperson earned a $200 commission which is 3% of a sale. How much was the sale? ______

3. Robin needs to make a car payment of $215.60 next week. What is the amount of sales she must make to earn a commission to cover this payment if the rate of commission is $9\frac{1}{2}$%? ______

4. Joan Yong must pay state income tax. Her earnings are $1,500 for the month. How much state income tax must be deducted from her gross earnings if the tax is $8 plus .8% of her excess income over $1,000? ______

5. How much will Jan Carter earn on $3,200 sales at a commission rate of $7\frac{1}{4}$%? ______

6. Sally's income was $2,448 last month. Sales were $20,400. What rate of commission did she earn? ______

7. Joan earns $500 per month plus 4% on annual sales. Jim earns a straight 6.5% commission on all sales.
   a. If total sales for both were $245,000, who earned the greater annual income?

   b. By how much?

ADVANCED APPLICATIONS

Solve the following problems using the information given.

1. Brett Wagner is paid a salary of $750 per month. In addition, he receives a graduated commission of 2% on sales of $2,000 through $4,000; 4% on sales over $4,000 through $6,500; and 6% on all sales over $6,500. Sales for the first three months of the year were: January, $10,000; February, $6,595; and March, $9,240. Determine his gross salary for each month.
   a. January

   b. February

   c. March

2. The payroll register for Purell Company shows employees' total gross earnings of $23,671. If federal income tax deductions total $3,256, what is the federal tax deposit for the month?

3. A salesperson earns $220 a week in salary plus a 4% commission on all sales. How much must he sell each week to earn a total of $350?

The Hodgepodge Shop

series III A Retail Application

"Life is hard by the yard;
by the inch, life's a cinch"

Welcome back to the Hodgepodge Shop. You will continue in the training program by working with accounts payable and payroll. Remember, it is not always the "smartest" individual who is hired, but the person who possesses the work ethic characteristics necessary to move up the ladder of success.

After completing the accounts payable and payroll training program, you should be able to:

1. calculate trade and cash discounts on invoices
2. write checks to pay creditors and complete check stubs
3. complete a payroll register
4. maintain check stubs for payment of payroll
5. complete the federal tax return

TERMINOLOGY

The following terminology is used in this application. Become familiar with the meaning of each term so that you understand its usage.

1. **Accounts Payable Report**—a report showing all information needed to pay a creditor.
2. **Creditor**—a company/person who is owed money.
3. **Employer's Quarterly Federal Tax Return**—a form showing the amount of tax deposited during a quarter.

ON-THE-JOB TRAINING

In this segment of your training, you will be calculating the amount The Hodgepodge Shop owes for merchandise purchased and will be writing checks to pay the accounts payable. You will then work with payroll to complete the payroll register and check stubs for the payment of the payroll. You will also complete a federal tax return. Read each job assignment carefully. Follow each step outlined for you. Do not skip any steps.

Good luck! Remember, you are a trainee, so ask for help along the way as you need it.

Assignment 1: Invoices and Checks

The accountant wants you to pay the invoices due. A clerk has verified the invoices and prepared an **Accounts Payable Report**. The report shows the company name, invoice number, invoice total, trade and cash discounts, and date of invoice. You are to complete the information on the report before writing checks to pay the invoices.

March 1

1. Remove the Accounts Payable Report (Form 1, page 237). Calculate the amount due for each invoice. Use March 1 as today's date to determine if a cash discount is allowed. Add any freight charge to obtain the net amount due.
2. Remove the checks and check stubs (Forms 2 through 7). Write one check to each company on the Accounts Payable Report. Your supervisor will sign the checks. Keep the check stubs current. Write the invoice number or numbers on the check stub. Use the numbered checks and stubs beginning with 577. The beginning balance is $10,015.62.
3. Verify the checks by totaling the amounts of the checks and comparing the sum with the total of the Accounts Payable Report. Submit the Accounts Payable Report to your supervisor. The report will be sent to data entry for updating each **creditor's** account.
4. Detach the checks and submit them to your supervisor. File the stubs for use in Assignment 3.

Assignment 2: Payroll Register

Every Friday a computer printout of a payroll register is received from data entry. It shows the employees' names, exemptions, marital status, hours worked, and hourly wage. All hours worked over 40 are calculated at time and a half. Each employee is covered by a health insurance policy paid by The Hodgepodge Shop. Each employee has the option of including his or her family by having $32.40 deducted from weekly earnings.

March 2

1. Remove the Payroll Register (Form 8). Find the net pay of each employee by completing Steps 2 through 8. If you take advantage of all features on your calculator, net pay for each individual can be computed in one continuous operation. Be sure that all calculations are correct. * ***Remember, rate × 1.5 × number of overtime hours = overtime pay.***
2. Calculate regular earnings for each employee. Accumulate in GT or Memory.

3. Calculate and record overtime earnings for each employee. Accumulate in GT or Memory.
4. Calculate and record FICA tax using 7.65%. Use each employee's total from Step 2 and multiply by .0765. Subtract this deduction from GT or Memory.
5. Refer to the Federal Tax Table on page 227 in Unit 13. Remove the State Tax Table (Form 9). Use the two tables to record the federal and state taxes on the payroll register. File the tax tables after recording the taxes. Subtract these deductions from GT or Memory.
6. Record family insurance payments for individuals with an asterisk in the insurance column. Subtract this deduction from GT or Memory.
7. Calculate and record net pay. Total GT or Memory.
8. Find the total gross earnings, FICA tax, federal tax, state tax, insurance, and net pay. Verify the accuracy of the payroll register.

Assignment 3: Payroll Checks

The payroll checks are to be prepared by hand for each of the employees listed on the payroll register (Form 8). You are to record the correct information on the check stubs.

March 2

1. Remove the check stubs (Forms 10 and 11).
2. Bring the balance forward from check stub #582 (in this Series). Add a deposit of $3,500. Continue to use the check stubs in numerical order, beginning with #583.
3. Correctly fill in the stubs just as if you were writing the checks. Be sure each employee's name is written correctly and the correct balance carried forward to each stub.
4. Add a deposit of $11,200 on check stub #606.
5. Submit the payroll register (Form 8) to your supervisor.

Assignment 4: Federal Tax Return

The taxes deducted from the employees' wages must be paid to the proper government agency. The Hodgepodge Shop makes a monthly deposit at its bank to pay federal income taxes and social security (FICA) taxes due. A report is sent to the IRS quarterly to show the amount of taxes deposited with the bank, the federal withholding taxes deducted from the employees' wages, and the total FICA taxes (both those deducted from the employees' wages and the employer's contribution).

April 1

Remove the **Employer's Quarterly Federal Tax Return** (Form 12) and follow these steps to complete the report. * ***Some of the lines do not apply and are to be left blank.***

1. **Line 1a**—record the number of employees for the pay period that includes March 12th. The number of employees is 23.

2. **Line 2**—record the total wages earned by all the employees. The amount is $103,839.84. This amount was found by totaling all the wages for each week during the quarter.
3. **Line 3**—record the total amount of federal taxes withheld for all the employees during the quarter. The amount is $13,238.40.
4. **Line 5**—record the amount of federal taxes to be paid to the federal government. Since no amount was recorded on line 4, copy the figure from line 3. The amount written should be $13,238.40.
6. **Line 6**—since all wages recorded on line 2 are taxable social security wages, record the amount from line 2 after the dollar sign. Multiply the amount by 15.3% and record the amount on line 6. * ***Notice that the employer must pay one half of the total social security taxes due on taxable wages.***
7. **Lines 8 and 10**—copy the amount from line 6 on lines 8 and 10.
8. **Lines 14 and 16**—add the amounts on lines 5 and 10 and record on lines 14 and 16.
9. **Line 17**—record $28,995.30. This is the amount of money already deposited in the bank during the first quarter.
10. **Line 18**—subtract line 17 from 16 and record on line 18. This is the amount that will be sent to the federal government with Form 941.
11. A check for the amount on line 18 on Form 941 is written to Federal Farm Union Bank. Fill in check stub #606 to show payment of taxes due. Federal Farm Union Bank will be given your check along with a *Federal Tax Deposit Coupon*. This coupon, completed by your supervisor, will direct the bank to deposit your taxes due with the Internal Revenue Service.
12. Submit the Employer's Quarterly Federal Tax Return and check stubs to your supervisor. Your supervisor will complete any other appropriate portions of the form.

Form 1

ACCOUNTS PAYABLE REPORT

March 1

Invoice Number	Company	Date of Invoice	Amount of Invoice	Trade Discount	Amount Due After Trade Discount	Cash Discount	Amount Due After Cash Discount	Freight Charge	Amount Due
1. 077623	Goodman's Gift House	2-27	$1,420.00	10-20		2/10,n/30		$15.00	
2. 61322	Accessories & More	2-20	576.30	15-5-10		3/10,n/30		5.00	
3. 220266	Southern Stationery Supply Co.	2-16	51.40	10-10-10		n/30		None	
4. 47786	S.R. Paper Products	2-16	182.52	5-10-5		2/15,n/60		None	
5. 077845	Goodman's Gift House	2-25	284.97	10-10		2/10,n/30		3.00	
6. 9822	Runsey's Supply House	2-17	178.00	12-5-5		4/15,n/60		None	
7. 47901	S.R. Paper Products	2-12	66.23	5-10-5		2/15,n/60		None	
8. 164091	Frames, Inc.	2-21	1,075.00	15-10-5		3/15,n/60		7.00	
9. 080042	Goodman's Gift House	2-28	563.80	10-10		2/10,n/30		6.00	
10. 226981	Southern Stationery Supply Co.	2-9	439.60	10-10-10		n/30		None	
								Total	

Form 2

CHECK **577**

______ 19 ______

TO ______

FOR ______

	DOLLARS	CENTS
Balance Brought Forward		
Amt. Deposited		
TOTAL		
Amt. this Check		
Balance Carried Forward		

The Hodgepodge Shop
705 4th Ave. S.
Minneapolis, MN 55415-2232

No. **577**

______ 19 ______ 17-05/910

PAY TO THE ORDER OF ______ $ ______

______ DOLLARS

FOR CLASSROOM USE ONLY

FEDERAL FARM UNION BANK
MINNEAPOLIS, MN 55415-2056

FOR ______

⑆091000501⑆ 577⑈5444302⑈

Form 3

CHECK **578**

______ 19 ______

TO ______

FOR ______

	DOLLARS	CENTS
Balance Brought Forward		
Amt. Deposited		
TOTAL		
Amt. this Check		
Balance Carried Forward		

The Hodgepodge Shop
705 4th Ave. S.
Minneapolis, MN 55415-2232

No. **578**

______ 19 ______ 17-05/910

PAY TO THE ORDER OF ______ $ ______

______ DOLLARS

FOR CLASSROOM USE ONLY

FEDERAL FARM UNION BANK
MINNEAPOLIS, MN 55415-2056

FOR ______

⑆091000501⑆ 578⑈5444302⑈

Form 4

CHECK **579**

______ 19 ______

TO ______

FOR ______

	DOLLARS	CENTS
Balance Brought Forward		
Amt. Deposited		
TOTAL		
Amt. this Check		
Balance Carried Forward		

The Hodgepodge Shop
705 4th Ave. S.
Minneapolis, MN 55415-2232

No. **579**

______ 19 ______ 17-05/910

PAY TO THE ORDER OF ______ $ ______

______ DOLLARS

FOR CLASSROOM USE ONLY

FEDERAL FARM UNION BANK
MINNEAPOLIS, MN 55415-2056

FOR ______

⑆091000501⑆ 579⑈5444302⑈

Form 5

CHECK **580**

______ 19 ____
TO ______
FOR ______

	DOLLARS	CENTS
Balance Brought Forward		
Amt. Deposited		
TOTAL		
Amt. this Check		
Balance Carried Forward		

The Hodgepodge Shop
705 4th Ave. S.
Minneapolis, MN 55415-2232

No. **580**

______ 19 ____ 17-05/910

PAY TO THE ORDER OF ______ $ ______

______ DOLLARS

FOR CLASSROOM USE ONLY

FEDERAL FARM UNION BANK
MINNEAPOLIS, MN 55415-2056

FOR ______

⑆091005001⑆ 580⑈ 5443 02⑈

Form 6

CHECK **581**

______ 19 ____
TO ______
FOR ______

	DOLLARS	CENTS
Balance Brought Forward		
Amt. Deposited		
TOTAL		
Amt. this Check		
Balance Carried Forward		

The Hodgepodge Shop
705 4th Ave. S.
Minneapolis, MN 55415-2232

No. **581**

______ 19 ____ 17-05/910

PAY TO THE ORDER OF ______ $ ______

______ DOLLARS

FOR CLASSROOM USE ONLY

FEDERAL FARM UNION BANK
MINNEAPOLIS, MN 55415-2056

FOR ______

⑆091005001⑆ 581⑈ 5443 02⑈

Form 7

CHECK **582**

______ 19 ____
TO ______
FOR ______

	DOLLARS	CENTS
Balance Brought Forward		
Amt. Deposited		
TOTAL		
Amt. this Check		
Balance Carried Forward		

The Hodgepodge Shop
705 4th Ave. S.
Minneapolis, MN 55415-2232

No. **582**

______ 19 ____ 17-05/910

PAY TO THE ORDER OF ______ $ ______

______ DOLLARS

FOR CLASSROOM USE ONLY

FEDERAL FARM UNION BANK
MINNEAPOLIS, MN 55415-2056

FOR ______

⑆091005001⑆ 582⑈ 5443 02⑈

THE HODGEPODGE SHOP
Payroll Register
March 2

Name	Marital Status	Exemp-tions	Gross Earnings					Deductions				Total Net Pay
			Hours	Hourly Wage	Regular Wage	OT Wage	Total	FICA Tax	Federal Tax	State Tax	Insurance	
1. Ainsworth, A.	S	1	40	$ 8.50								
2. Ashley, T.	S	2	44	8.75								
3. Ballew, C.	M	5	40	8.75							*	
4. Colwell, D.	M	3	40	8.75							*	
5. Compton, L.	M	2	39.2	9.25								
6. Departie, C.	S	1	36	9.25								
7. Epley, R.	M	4	48	10.00							*	
8. Garza, R.	M	3	40	8.50								
9. Hensley, D.	S	2	42.5	8.75								
10. Killough, M.	S	3	38	8.75							*	
11. Lowman, J.	M	2	43	8.75							*	
12. McKinley, H.	M	5	45	9.25							*	

Name	Marital Status	Exemp-tions	Gross Earnings					Deductions				Total Net Pay
			Hours	Hourly Wage	Regular Wage	OT Wage	Total	FICA Tax	Federal Tax	State Tax	Insurance	
13. Mills, C.	M	4	40	$ 9.00							*	
14. Mullin, T.	M	2	36	9.50								
15. Pasley, B.	S	1	45.5	8.50								
16. Richter, T.	M	3	40	8.50							*	
17. Saller, D. R.	S	1	46	8.50								
18. Savage, G.	S	2	48.4	9.00							*	
19. Swinhart, S.	M	2	39.6	8.75							*	
20. Taylor, B.	M	3	40	8.75							*	
21. Thurman, J.	S	1	40	10.00								
22. Turner, M.	M	3	42.8	8.75							*	
23. Yaeger, D.	M	2	43	8.75								
						Totals						

MINNESOTA INCOME TAX WITHHOLDING TABLE

Form 9

Single Employees
PAID ONCE A WEEK

If the employee's wages are at least	but less than	0	1	2	3	4	5	6	7	8	9	10 or more
		the amount to withhold is *(all amounts shown are in whole dollars)* —										
$330	$340	20	17	14	12	10	7	5	3	0	0	0
340	350	21	18	15	13	10	8	6	3	1	0	0
350	360	22	19	16	13	11	8	6	4	2	0	0
360	370	23	19	16	14	11	9	7	4	2	0	0
370	380	23	20	17	14	12	10	7	5	3	0	0
380	390	24	21	18	15	13	10	8	6	3	1	0
390	400	25	22	19	16	13	11	9	6	4	2	0
400	410	26	23	20	16	14	11	9	7	5	2	0
410	420	27	23	20	17	14	12	10	7	5	3	1
420	430	27	24	21	18	15	13	10	8	6	3	1
430	440	28	25	22	19	16	13	11	9	6	4	2
440	450	29	26	23	20	17	14	12	9	7	5	2
450	460	30	27	24	20	17	14	12	10	8	5	3
460	470	31	27	24	21	18	15	13	10	8	6	4
470	480	31	28	25	22	19	16	13	11	9	6	4
480	490	32	29	26	23	20	17	14	12	9	7	5
490	500	33	30	27	24	21	18	15	12	10	8	5
500	510	34	31	28	24	21	18	15	13	11	8	6
510	520	35	31	28	25	22	19	16	13	11	9	7
520	530	35	32	29	26	23	20	17	14	12	9	7
530	540	36	33	30	27	24	21	18	15	12	10	8
540	550	37	34	31	28	25	22	18	15	13	11	8
550	560	38	35	32	28	25	22	19	16	14	11	9
560	570	39	35	32	29	26	23	20	17	14	12	10
570	580	39	36	33	30	27	24	21	18	15	12	10
580	590	40	37	34	31	28	25	22	19	15	13	11
590	600	41	38	35	32	29	26	22	19	16	14	11
600	610	42	39	36	32	29	26	23	20	17	14	12
610	620	43	39	36	33	30	27	24	21	18	15	13
620	630	43	40	37	34	31	28	25	22	19	16	13
630	640	44	41	38	35	32	29	26	23	19	16	14
640	650	45	42	39	36	33	30	26	23	20	17	14
650	660	46	43	40	36	33	30	27	24	21	18	15
660	670	47	43	40	37	34	31	28	25	22	19	16
670	680	47	44	41	38	35	32	29	26	23	20	17
$680 and over		8.0 percent of the excess over $680 plus *(round total to nearest whole dollar)*										
		48	45	42	38	35	32	29	26	23	20	17

MINNESOTA INCOME TAX WITHHOLDING TABLE

Married Employees

PAID ONCE A WEEK

If the employee's wages are		and the number of withholding allowances is										
at least	but less than	0	1	2	3	4	5	6	7	8	9	10 or more
		the amount to withhold is *(all amounts shown are in whole dollars)* —										
$330	$340	16	14	12	9	7	5	3	0	0	0	0
340	350	17	15	12	10	8	5	3	1	0	0	0
350	360	18	15	13	11	8	6	4	1	0	0	0
360	370	18	16	14	11	9	7	4	2	0	0	0
370	380	19	17	14	12	10	7	5	3	0	0	0
380	390	19	17	15	12	10	8	6	3	1	0	0
390	400	20	18	15	13	11	8	6	4	2	0	0
400	410	21	18	16	14	11	9	7	4	2	0	0
410	420	21	19	17	14	12	10	7	5	3	0	0
420	430	22	20	17	15	13	10	8	6	3	1	0
430	440	23	20	18	15	13	11	9	6	4	2	0
440	450	23	21	18	16	14	11	9	7	5	2	0
450	460	24	21	19	17	14	12	10	7	5	3	1
460	470	25	22	20	17	15	13	10	8	6	3	1
470	480	26	23	20	18	16	13	11	9	6	4	2
480	490	27	23	21	18	16	14	12	9	7	5	2
490	500	27	24	21	19	17	14	12	10	8	5	3
500	510	28	25	22	20	17	15	13	10	8	6	4
510	520	29	26	23	20	18	16	13	11	9	6	4
520	530	30	27	24	21	19	16	14	12	9	7	5
530	540	31	27	24	21	19	17	15	12	10	8	5
540	550	31	28	25	22	20	17	15	13	11	8	6
550	560	32	29	26	23	20	18	16	13	11	9	7
560	570	33	30	27	24	21	19	16	14	12	9	7
570	580	34	31	28	25	22	19	17	15	12	10	8
580	590	35	31	28	25	22	20	18	15	13	11	8
590	600	35	32	29	26	23	20	18	16	14	11	9
600	610	36	33	30	27	24	21	19	16	14	12	10
610	620	37	34	31	28	25	22	19	17	15	12	10
620	630	38	35	32	29	25	22	20	18	15	13	11
630	640	39	35	32	29	26	23	21	18	16	14	11
640	650	39	36	33	30	27	24	21	19	17	14	12
650	660	40	37	34	31	28	25	22	19	17	15	13
660	670	41	38	35	32	29	26	23	20	18	15	13
670	680	42	39	36	33	29	26	23	21	18	16	14
$680 and over		8.0 percent of the excess over $680 plus *(round total to nearest whole dollar)*										
		42	39	36	33	30	27	24	21	19	16	14

Form 10

CHECK 583

19 ______

TO ______

FOR ______

	DOLLARS	CENTS
Balance Brought Forward		
Amt. Deposited		
TOTAL		
Amt. this Check		
Balance Carried Forward		

CHECK 584

19 ______

TO ______

FOR ______

	DOLLARS	CENTS
Balance Brought Forward		
Amt. Deposited		
TOTAL		
Amt. this Check		
Balance Carried Forward		

CHECK 585

19 ______

TO ______

FOR ______

	DOLLARS	CENTS
Balance Brought Forward		
Amt. Deposited		
TOTAL		
Amt. this Check		
Balance Carried Forward		

CHECK 586

19 ______

TO ______

FOR ______

	DOLLARS	CENTS
Balance Brought Forward		
Amt. Deposited		
TOTAL		
Amt. this Check		
Balance Carried Forward		

CHECK 587

19 ______

TO ______

FOR ______

	DOLLARS	CENTS
Balance Brought Forward		
Amt. Deposited		
TOTAL		
Amt. this Check		
Balance Carried Forward		

CHECK 588

19 ______

TO ______

FOR ______

	DOLLARS	CENTS
Balance Brought Forward		
Amt. Deposited		
TOTAL		
Amt. this Check		
Balance Carried Forward		

CHECK 589

19 ______

TO ______

FOR ______

	DOLLARS	CENTS
Balance Brought Forward		
Amt. Deposited		
TOTAL		
Amt. this Check		
Balance Carried Forward		

CHECK 590

19 ______

TO ______

FOR ______

	DOLLARS	CENTS
Balance Brought Forward		
Amt. Deposited		
TOTAL		
Amt. this Check		
Balance Carried Forward		

CHECK 591

19 ______

TO ______

FOR ______

	DOLLARS	CENTS
Balance Brought Forward		
Amt. Deposited		
TOTAL		
Amt. this Check		
Balance Carried Forward		

CHECK 592

19 ______

TO ______

FOR ______

	DOLLARS	CENTS
Balance Brought Forward		
Amt. Deposited		
TOTAL		
Amt. this Check		
Balance Carried Forward		

CHECK 593

19 ______

TO ______

FOR ______

	DOLLARS	CENTS
Balance Brought Forward		
Amt. Deposited		
TOTAL		
Amt. this Check		
Balance Carried Forward		

CHECK 594

19 ______

TO ______

FOR ______

	DOLLARS	CENTS
Balance Brought Forward		
Amt. Deposited		
TOTAL		
Amt. this Check		
Balance Carried Forward		

Form 11

CHECK **595**

_______________ 19 _____

TO _______________

FOR _______________

	DOLLARS	CENTS
Balance Brought Forward		
Amt. Deposited		
TOTAL		
Amt. this Check		
Balance Carried Forward		

CHECK **596**

_______________ 19 _____

TO _______________

FOR _______________

	DOLLARS	CENTS
Balance Brought Forward		
Amt. Deposited		
TOTAL		
Amt. this Check		
Balance Carried Forward		

CHECK **597**

_______________ 19 _____

TO _______________

FOR _______________

	DOLLARS	CENTS
Balance Brought Forward		
Amt. Deposited		
TOTAL		
Amt. this Check		
Balance Carried Forward		

CHECK **598**

_______________ 19 _____

TO _______________

FOR _______________

	DOLLARS	CENTS
Balance Brought Forward		
Amt. Deposited		
TOTAL		
Amt. this Check		
Balance Carried Forward		

CHECK **599**

_______________ 19 _____

TO _______________

FOR _______________

	DOLLARS	CENTS
Balance Brought Forward		
Amt. Deposited		
TOTAL		
Amt. this Check		
Balance Carried Forward		

CHECK **600**

_______________ 19 _____

TO _______________

FOR _______________

	DOLLARS	CENTS
Balance Brought Forward		
Amt. Deposited		
TOTAL		
Amt. this Check		
Balance Carried Forward		

CHECK **601**

_______________ 19 _____

TO _______________

FOR _______________

	DOLLARS	CENTS
Balance Brought Forward		
Amt. Deposited		
TOTAL		
Amt. this Check		
Balance Carried Forward		

CHECK **602**

_______________ 19 _____

TO _______________

FOR _______________

	DOLLARS	CENTS
Balance Brought Forward		
Amt. Deposited		
TOTAL		
Amt. this Check		
Balance Carried Forward		

CHECK **603**

_______________ 19 _____

TO _______________

FOR _______________

	DOLLARS	CENTS
Balance Brought Forward		
Amt. Deposited		
TOTAL		
Amt. this Check		
Balance Carried Forward		

CHECK **604**

_______________ 19 _____

TO _______________

FOR _______________

	DOLLARS	CENTS
Balance Brought Forward		
Amt. Deposited		
TOTAL		
Amt. this Check		
Balance Carried Forward		

CHECK **605**

_______________ 19 _____

TO _______________

FOR _______________

	DOLLARS	CENTS
Balance Brought Forward		
Amt. Deposited		
TOTAL		
Amt. this Check		
Balance Carried Forward		

CHECK **606**

_______________ 19 _____

TO _______________

FOR _______________

	DOLLARS	CENTS
Balance Brought Forward		
Amt. Deposited		
TOTAL		
Amt. this Check		
Balance Carried Forward		

Form **941** | Department of the Treasury Internal Revenue Service | 4242

Employer's Quarterly Federal Tax Return

Form 12

See Circular E for more information concerning Employment Tax Returns.

Please type or print.

OMB No. 1545-0029
Expires: 5-31-91

Your name, address, employer identification number, and calendar quarter of return. (If not correct, please change.)

The Hodgepodge Shop
705 4th Ave. S.
Minneapolis, MN 55415-2232

Date quarter ended
3/31/--

Employer identification number
10-0000-000

T
FF
FD
FP
I
T

IRS Use

1 1 1 1 1 1 1 1 1 1 2 3 3 3 3 3 3 4 4 4

5 5 5 6 7 8 8 8 8 8 8 9 9 9 10 10 10 10 10 10 10 10 10 10

If address is different from prior return, check here ☐

If you do not have to file returns in the future, check here . . . ☐ Date final wages paid . . . .

If you are a seasonal employer, see **Seasonal employer** on page 2 and check here . . . ☐

1a	Number of employees (except household) employed in the pay period that includes March 12th	1a	
b	If you are a subsidiary corporation AND your parent corporation files a consolidated Form 1120, enter parent corporation employer identification number (EIN) . . ▶ **1b** –		
2	Total wages and tips subject to withholding, plus other compensation . . . ▶	2	
3	Total income tax withheld from wages, tips, pensions, annuities, sick pay, gambling, etc. . ▶	3	
4	Adjustment of withheld income tax for preceding quarters of calendar year (see instructions) . . ▶	4	
5	Adjusted total of income tax withheld (line 3 as adjusted by line 4—see instructions) . . .	5	
6	Taxable social security wages paid . . . $ ________ × 15.3% (.153) . .	6	
7a	Taxable tips reported . . . $ ________ × 15.3% (.153) . .	7a	
b	Taxable hospital insurance wages paid . . . $ ________ × 2.9% (.029) . .	7b	
8	Total social security taxes (add lines 6, 7a, and 7b) . . .	8	
9	Adjustment of social security taxes (see instructions for required explanation) . . .	9	
10	Adjusted total of social security taxes (line 8 as adjusted by line 9—see instructions) . . ▶	10	
11	Backup withholding (see instructions) . . .	11	
12	Adjustment of backup withholding tax for preceding quarters of calendar year . . ▶	12	
13	Adjusted total of backup withholding (line 11 as adjusted by line 12) . . .	13	
14	Total taxes (add lines 5, 10, and 13) . . .	14	
15	Advance earned income credit (EIC) payments, if any . . . ▶	15	
16	Net taxes (subtract line 15 from line 14). **This must equal line IV below** (plus line IV of Schedule A (Form 941) if you have treated backup withholding as a separate liability) . . .	16	
17	Total deposits for quarter, including overpayment applied from a prior quarter, from your records . ▶	17	
18	Balance due (subtract line 17 from line 16). This should be less than $500. Pay to IRS . . ▶	18	

19 If line 17 is more than line 16, enter overpayment here ▶ $ ________ and check if to be: ☐ Applied to next return **OR** ☐ Refunded.

Record of Federal Tax Liability (Complete if line 16 is $500 or more.) See the instructions on page 4 for details before checking these boxes.

Check only if you made eighth-monthly deposits using the 95% rule ▶ ☐ Check only if you are a first time 3-banking-day depositor ▶ ☐

Do NOT Show Federal Tax Deposits Here

Date wages paid	Show tax liability here, **not deposits.** IRS gets deposit data from FTD coupons.					
	First month of quarter		Second month of quarter		Third month of quarter	
1st through 3rd	A		I		Q	
4th through 7th	B		J		R	
8th through 11th	C		K		S	
12th through 15th	D		L		T	
16th through 19th	E		M		U	
20th through 22nd	F		N		V	
23rd through 25th	G		O		W	
26th through the last	H		P		X	
Total liability for month	*I*		*II*		*III*	

IV Total for quarter (add lines *I*, *II*, and *III*). **This must equal line 16 above** . . . ▶

Sign Here Under penalties of perjury, I declare that I have examined this return, including accompanying schedules and statements, and to the best of my knowledge and belief, it is true, correct, and complete.

Signature ▶ **Title** ▶ **Date** ▶

For Paperwork Reduction Act Notice, see page 2.

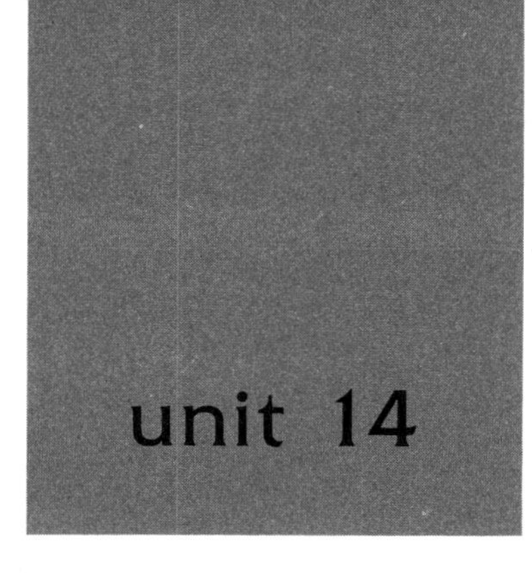

unit 14 Pricing

"Never put your wishbone where your backbone ought to be."

Businesses operate to make a profit. When merchandise is sold, the selling price must be high enough to cover the amount the business paid for the merchandise, plus the expenses of doing business—utilities, supplies, building costs, payroll—and the net profit. As a businessperson, you need to be able to determine a selling price that will cover cost, expenses, and profit.

After studying this unit, you should be able to:

1. understand the difference between markup based on cost and markup based on selling price
2. calculate markup based on *selling price* and determine the selling price
3. calculate markup based on *cost* and determine the selling price
4. solve business math problems using markup and markdown

MATH TERMS AND CONCEPTS

The following terminology is used in this unit. Become familiar with the meaning of each term so that you understand its usage as you develop math skills.

1. **Markdown**—an amount deducted from the selling price.
2. **Markup Amount or Markup**—an amount added to the wholesale cost.
3. **Markup Rate**—the percent that is multiplied times cost or selling price to find the markup amount.
4. **Overhead**—the costs involved in operating a business.
5. **Wholesale Cost**—the amount the retailer pays for merchandise, usually called *cost.*

Markup

When a retailer prices merchandise to sell, the selling price must include **wholesale cost**, **overhead**, and *net profit.* The sum of overhead and net profit equals the **markup amount**. The sum of cost and markup amount equals the *selling price.*

Overhead + Net Profit = Markup

Cost + Markup = Selling Price

Find each missing item.

	Cost	Markup	Selling Price
1.	$25	$17	__________
2.	__________	$32	$88.50
3.	$13	__________	$24
4.	$89	$27	__________
5.	$113	__________	$168

Markup is calculated either as a percent of the wholesale cost or as a percent of the selling price. Retailers usually base markup on selling price because they use the suggested retail price list given by the manufacturer. They also use total sales as a basis for comparing all areas of the business.

Manufacturers usually base markup on cost because they compare other areas of the business with inventory.

* ***When you compute markup, you are actually doing percentage, rate, and base problems like you did in Unit 8. See Unit 8 for a review.***

Markup Based on Cost

When markup is based on cost, the cost is always the base—100%. Let's look at the formulas and compare percentage and markup. Remember the triangle?

Percentage Formulas

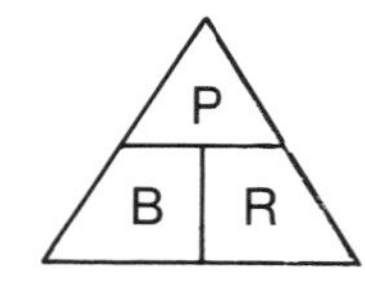

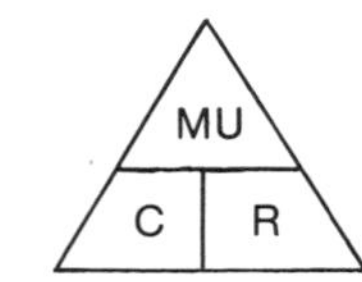

Markup Based on Cost

P = Markup = MU	MU = C × R
B = Cost = C	C = MU ÷ R
R = Rate = R	R = MU ÷ C

Example: A store pays $5 for a lawn chair and needs a 50% markup to cover overhead and net profit. The wholesale cost of the lawn chair is the base, the 50% gross profit is the **markup rate**, and the markup is the percentage. Find the markup amount and selling price.

C × R = MU
$5 × .5 = $2.50 = Markup Amount = Gross Profit

Cost + Markup = Selling Price
$5 + $2.50 = $7.50

Find the markup amount and the selling price for each item.

Use add-on percent.

Item	Wholesale Cost	Rate	Markup Amount	Selling Price
1.	$13.86	30%	______	______
2.	$41.00	25%	______	______
3.	$5.31	7%	______	______
4.	$97.16	17%	______	______
5.	$111.62	19%	______	______

In working markup problems, one of the elements will be missing. You must determine which element you need to find and then use the correct formula to solve for it.

Find the missing element and the selling price in each of the following problems. Round answers to two decimal places.

Use the percent key.

	Cost	Rate	Markup	Selling Price
6.	$90	15%	______	______
7.	$86	______	$53.75	______
8.	______	25%	$7.20	______
9.	$153.20	______	$84.26	______
10.	$720	60%	______	______

Markup Based on Selling Price

When markup is based on selling price, the selling price is always the base—100%. The markup amount and the rate are used the same way as in markup on cost. In the formulas, use SP to represent selling price.

Percentage Formulas

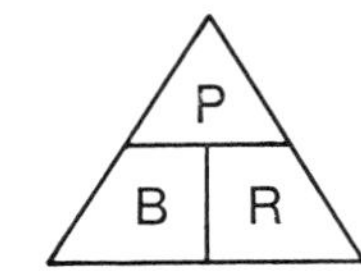

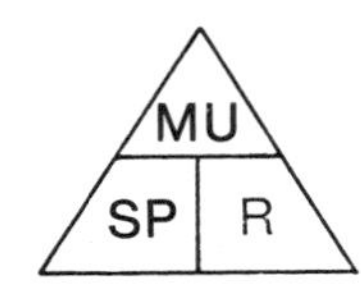

Markup Based on Selling Price

MU = SP × R

SP = MU ÷ R

R = MU ÷ SP

Example: A store sells an ice chest for $10. The markup is 40% of the selling price. Find the cost.

SP × R = MU
$10 × 40% = $4 (markup amount)

SP − MU = Cost
$10 − $4 = $6

Find the markup amount and the cost for each of the following problems.

Use the percent discount feature.

	Selling Price	Rate	Markup Amount	Cost
1.	$40	20%	________	________
2.	$364.80	45%	________	________
3.	$256.57	15%	________	________
4.	$180.91	63%	________	________
5.	$443.00	$22\frac{1}{2}$%	________	________

Again, in solving business problems, one of the elements besides markup will be missing. You must determine which element you need to find. Then use the correct formula to solve for it.

Solve the following problems. * ***Use your percent key.***

	Selling Price	Rate	Markup Amount	Cost
6.	$1,012	________	$723.58	________
7.	________	75%	$430.35	________

	Selling Price	Rate	Markup Amount	Cost
8.	________	20%	$26	________
9.	$22.66	________	$11.33	________
10.	$624	22.5%	________	________

Markdown

Businesses reduce the price of merchandise to compete with other stores, to attract customers, and to clear out discontinued items. A **markdown** is simply a reduction or discount on the original selling price.

Percentage Formulas

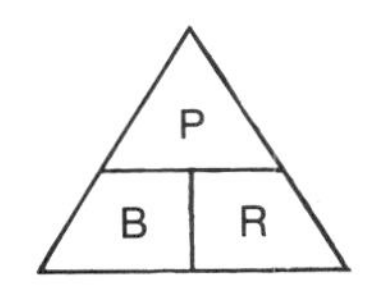

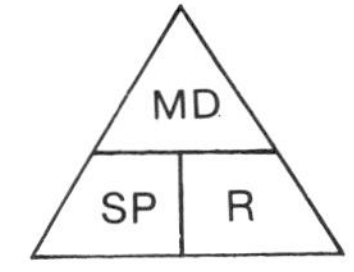

Markdown

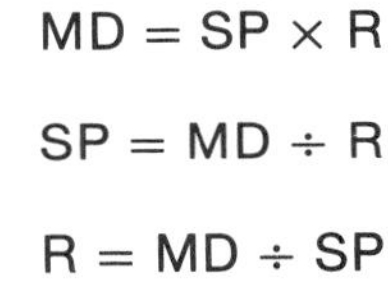

MD = SP × R

SP = MD ÷ R

R = MD ÷ SP

Example: Renee's Records advertises a tape player at 25% off the selling price of $40. This is a *markdown rate* of 25%.

SP × R = MD
$40 × .25 = $10 = Markdown Amount

SP − MD = RP
$40 − $10 = $30 = Reduced Price

Use the percent key to find the markdown amount. Operate the minus key to obtain the reduced price or use the memory to determine the reduced price. If your calculator has a −% key, use it.

Often it is possible to use aliquot parts when working with markdown. In the above example, it is simple to find $\frac{1}{4}$ of $40 and subtract.

$\frac{1}{4}$ of $40 is $10

$40 − $10 = $30

If the markdown is 20%, you can find 10%; then double the amount, and subtract mentally.

Sometimes a store may give the original selling price and the reduced price. You then need to calculate the rate. This again is a simple percentage problem.

Example: $40 = Selling Price
$10 = Markdown
? = Rate
$10 ÷ $40 = 25%

Solve the following problems for the markdown or the rate of markdown. Then find the reduced price.

	Selling Price	Markdown	Rate	Reduced Price
1.	$137.68	________	22%	________
2.	$89	$32.93	37%	________
3.	$391.07	$293.30	75%	________
4.	$13.80	$.69	5%	________
5.	$740.12	$355.26	48%	________

You can use the percent minus feature on your calculator.

Solving Business Problems

Before solving business problems, be sure the markup rate matches the cost or selling price. If the rate and percentage are given, they must match or represent the same thing.

Example: A dress costing $17 has a markup based on selling price of 32%. What is the selling price?

SP = ?
R = 32%
C = $17

In this problem, the markup is based on the missing element (selling price). If selling price equals 100% and markup is 32%, then cost equals 100 – 32%. A match is found by dividing the cost by the complement of the markup rate.

If your calculator has the MU/MD feature, use it to compute selling price. Refer to Appendix A for specific instructions.

100% – 32% = 68% (rate based on cost)
$17 ÷ 68% = $25 (selling price)

Example: A camera sells for $71.34 and has a markup of 23% based on cost. What did the camera cost?

SP = $71.34
R = 23%
C = ?

In this example, markup is based on the missing element (cost). If cost equals 100% and markup is 23%, then selling price will equal 123% (cost + markup = selling price).

You can use your MU/MD feature to compute cost if your calculator has this feature.

$71.34 ÷ 123% = $58 (cost)

Remember, selling price should be more than cost.

You can use a shortcut to compute cost or selling price if the markup amount is not needed. When you find the selling price only, you are finding a given percent *more than* the cost. Add the given percent to 100% before multiplying.

Example: A lawn chair costing \$5 will be marked up 50% based on cost.

100% + 50% = 150% = Rate
\$5 × 150% = \$7.50 = Selling Price

When you find the cost only, you are finding a given percent *less than* the selling price. Subtract the given percent from 100% before multiplying.

Example: An ice chest selling for \$10 has been marked up 40% based on selling price. Find the cost.

100% − 40% = 60% = Rate
\$10 × 60% = \$6 = Cost

* *Remember, use the percentage, base, and rate formulas to solve for markup. The percentage and rate must match!*

EVALUATING YOUR SKILLS

The following problems evaluate your skills in the material covered in Units 1 through 13. Some of the problems will evaluate your ability to work problems mentally while others will evaluate your ability to use your calculator. Work the problems as quickly as you can. Where necessary, use the metric conversion chart in Appendix B, page 339.

1. Write 17 thousandths as a decimal. __________
2. Write .063 as a percent. __________
3. Write 6% as a common fraction. __________
4. Convert 45 pounds to kilograms. __________
5. $\frac{1}{2}$ of $\frac{1}{3}$ = ? __________
6. How much is 3% of \$30? __________
7. Divide 1,000 by .1. __________
8. Find the common denominator: $\frac{4}{9} + \frac{3}{12} + \frac{4}{8} + \frac{3}{4}$ __________
9. Estimate the product: 872.14 × 34 __________

10. Write $\frac{5}{17}$ as a percent to the nearest tenth of a percent. __________

11. Change $\frac{1}{2}$% to a decimal. __________

12. Helen earns $9.25 an hour for a 40-hour week plus time and a half for overtime. Last week she worked 44 hours. What was her gross pay? __________

13. Sales for this year were $122,450. Last year's sales were $135,420. What was the percent increase or decrease to the nearest hundredth of a percent? __________

14. An $800 item offers a trade discount of 10, 20, 5. Find the cash price. __________

15. Lena Mueller received her bank statement and found a deposit for $125 was not recorded in her check register. A check for $32.50 was recorded as $23.50. A service charge in the amount of $4.50 was deducted from the bank statement. If her checkbook balance was $320.80, what is the adjusted balance? __________

APPLYING YOUR KNOWLEDGE

Solve the following problems. Be sure you read each problem carefully.

1. A store sells tires for $112 each. The markup rate on selling price is 37%.
   a. What is the markup amount? __________
   b. What is the cost? __________

2. A television has a markup of $75 which is 42% of selling price. What is the selling price? __________

3. A computer costs $1,357 and has a markup based on selling price of 72%. What is the selling price? __________

4. The Technology Company imports calculators from Japan. Each calculator sells for $88 and has a markup of $23 based on selling price. What is the markup rate?

____

5. A tuning fork costs $13 and has a markup based on cost of 14%. What is the markup amount?

____

6. A record sells for $12.95 and has a markup based on cost of 11%. What does the record cost?

____

7. A stereo costs $789 and has a markup based on cost of $233. What is the markup rate?

____

8. A pair of jeans has a markup of $9, which is a 25% markup based on cost. What is the cost?

____

9. What is the selling price if an item costs $178 and markup based on cost is 27%?

____

10. Merchandise is marked up 38% based on the selling price. What is the cost of the item if the amount of markup is $6?

____

11. A microwave oven costs $279 and sells for $450. What is the markup rate based on selling price?

____

12. Sound Systems recently purchased cassette players for $80.80 each. The markup on each cassette player is $20.20. What is the markup rate based on selling price?

____

13. A computer program sells for $120.30. The cost is $64.80. What is the markup rate based on cost?

____

14. A chair selling for $430 is marked down 30%. What is the new selling price?

____

15. A shirt selling for $30 is marked down to $27.60. What is the markdown rate?

____

ADVANCED APPLICATIONS

Using the information given, solve the following problems.

1. Determine the cost of a dress if the overhead and net profit are 35% of the selling price. The dress sells for $117. ______________

2. Determine the markup rate of an item that costs $478 and sells for $729 if markup is based on cost. ______________

3. An item has a list price of $274. The store receives a trade discount of 15% and 10%. What is the selling price if markup based on cost is 22%? ______________

4. Merchandise costing $395 is marked up 22% based on selling price. What is the selling price if it is marked down 10%? ______________

5. Merchandise lists for $113 with a trade discount of 15%. What is the selling price if the store wants a net profit of 9% and overhead is 21% of sales? ______________

6. A power lawn mower costs $152.50 and sells for $210.90. If overhead is $21.40, what is the net profit? ______________

7. Gardwild Furniture Store purchased lamps for $723 including freight charges of $56. If overhead and net profit are 27% of sales and a 5% discount is given to customers, what is the selling price? ______________

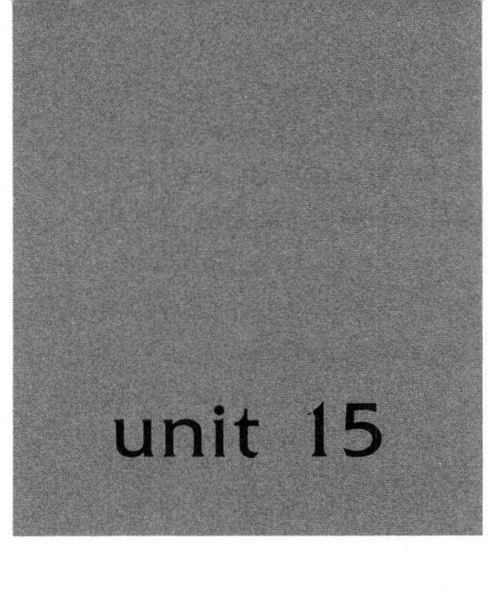

unit 15 Depreciation

"People are also known by the company they avoid."

Businesses buy equipment, machinery, and buildings to use in the operation of the business. These are called **assets**. As the assets are used, their value decreases or depreciates. Several methods are used to calculate how much an asset depreciates over a given period of time. **Depreciation** is considered an expense of doing business. The expense may be deducted from a business' profits when income tax returns are filed.

After studying this unit, you should be able to:

1. calculate depreciation using the straight-line method, the declining-balance method, and the sum-of-the-years-digits method
2. determine salvage value
3. determine a rate of depreciation based on the straight-line method
4. calculate book value
5. determine gain or loss on sale of an asset
6. solve business problems involving depreciation

MATH TERMS AND CONCEPTS

The following terminology is used in this unit. Become familiar with the meaning of each term so that you understand its usage as you develop math skills.

1. **Accumulated Depreciation**—the total amount of depreciation recorded during an asset's useful life.
2. **Accumulated Depreciation Rate**—the rate used to find accumulated sum-of-the-years-digits depreciation at any time during an asset's life.
3. **Annual Depreciation**—the amount of depreciation recorded for a year.
4. **Asset**—anything of value owned by a business.
5. **Book Value**—the original cost of an asset minus accumulated depreciation.
6. **Cost**—purchase price of an asset.
7. **Depreciation**—the loss of value of an asset through use and time.
8. **Gain on Sale**—when an item is disposed of for more than the book value.

9. **Loss on Sale**—when an item is disposed of for less than the book value.
10. **Salvage Value or Trade-In Value**—an asset's worth at the end of its useful life.
11. **Useful life**—the length of time an asset can be used.

Depreciation

Calculating **depreciation** involves three elements: the **cost** of the **asset**, the asset's **useful life**, and the **salvage value** of the asset. Both useful life and salvage value are estimates made by the company. Salvage value may be zero. The Internal Revenue Service publishes guidelines that companies can use to make estimates of useful life and salvage value. The type of depreciation used will affect net profit during the life of an asset.

Depreciable Amount

For some depreciation methods, you need to know the *depreciable amount*—the amount that will be depreciated. The depreciable amount is the difference between the cost (purchase price) and the salvage value.

Depreciable Amount = Purchase Price – Salvage Value

Example: Dr. Jenkins purchased a copy machine for $2,000. The life of the copier is estimated to be five years. The salvage value is estimated to be $500. The depreciable amount is $1,500.

$2,000 – $500 = $1,500 (depreciable amount)

Find the depreciable amount in each of the following problems.

	Purchase Price	Salvage Value	Depreciable Amount
1.	$8,000	$1,350	________
2.	$17,420	$3,600	________
3.	$980	-0-	________
4.	$7,350	$600	________
5.	$26,540	$10,000	________

Straight-Line Method

When a business uses *straight-line depreciation*, the asset depreciates the same amount each year. Once the depreciable amount is determined, divide the depreciation amount by the number of years in the asset's useful life.

Example: A truck is purchased for $13,000. The salvage value is estimated to be $3,000 at the end of five years. The depreciable amount would be $10,000 and the **annual depreciation** would be $2,000.

$13,000 – $3,000 = $10,000 (depreciable amount)
$10,000 ÷ 5 = $2,000 (depreciation per year)

Find the straight-line depreciation in each of the following problems.

	Purchase Price	Life	Salvage Value	Annual Depreciation
1.	$7,000	3 years	$450	________
2.	$25,750	12 years	$3,245	________
3.	$3,000	5 years	$350	________
4.	$12,000	6 years	$2,000	________
5.	$6,000	5 years	$500	________

Book Value

Book value is the value of an asset after **accumulated depreciation** is subtracted from the original cost. The salvage value is not considered.

Book Value = Purchase Price − Accumulated Depreciation

Example: An automobile purchased by the company for $12,320 is depreciated $1,500 a year for five years. At the end of three years the book value would be:

$1,500 × 3 = $4,500 (accumulated depreciation)
$12,320 − $4,500 = $7,820 (book value)

For each of the following, find the book value at the end of the first, second, and third years.

Use repeat subtraction.

	Cost	Annual Depreciation	1st Year	2nd Year	3rd Year
1.	$16,800	$1,300	________	________	________
2.	$8,700	$900	________	________	________
3.	$92,000	$9,000	________	________	________

	Cost	Annual Depreciation	1st Year	2nd Year	3rd Year
4.	$6,400	$750	________	________	________
5.	$39,500	$5,600	________	________	________

Rate of Depreciation

The *rate of depreciation* can be used in computing straight-line depreciation. The rate is based on the useful life of an asset. The useful life becomes the denominator of a fraction with 1 as the numerator.

Example: An asset is estimated to have a four-year life. Based on the straight-line method, the asset depreciates $\frac{1}{4}$ or 25% of the depreciable amount each year. An asset estimated to have a five-year life would have a rate of $\frac{1}{5}$ or 20%.

8-year estimated life = $\frac{1}{8}$ = 12.5% annual depreciation rate
10-year estimated life = $\frac{1}{10}$ = 10% annual depreciation rate
30-year estimated life = $\frac{1}{30}$ = $3\frac{1}{3}$% annual depreciation rate

The rate of depreciation is usually stated as a percent rather than as a fraction.

Find the rate of depreciation based on the straight-line method for each of the following problems.

Remember Aliquot Parts!

	Useful Life	Rate of Depreciation: Fraction	Rate of Depreciation: Percent
1.	4	________	________
2.	7	________	________
3.	6	________	________
4.	12	________	________
5.	5	________	________

Declining-Balance Method

When a business uses *declining-balance depreciation,* the asset depreciates more during the first years of ownership. As the asset nears the end of its useful life, the annual depreciation decreases. The depreciable amount is the same as the purchase price because salvage value is *not* considered. However, the book value *cannot* be less than the estimated salvage value. The rate of depreciation is always used in the declining-balance method.

Declining-balance depreciation is calculated by first multiplying the straight-line rate of depreciation times two. The yearly depreciation is then found by multiplying the *book value* by the depreciation rate.

Example: A truck purchased for $13,000 has an estimated life of five years. It has an estimated salvage value of $3,000.

$\frac{1}{5}$ = 20%

20% × 2 = 40% (rate of depreciation)

$13,000 × 40% = $5,200 (first year depreciation)
$13,000 − $5,200 = $7,800 (book value end of first year)

$7,800 × 40% = $3,120 (second year depreciation)
$7,800 − $3,120 = $4,680 (book value end of second year)

$4,680 × 40% = $1,872 (third year depreciation)
$4,680 − $1,872 = $2,808 (This brings the book value $192 below the salvage value of $3,000).
$3,000 − $2,808 = $192

$4,680 − $3,000 = $1,680 (actual third year depreciation)

Remember, book value cannot be less than salvage value.

Although the depreciation rate remained the same, the amount of depreciation became less each year. This is because the book value declined each year.

In the following problems, find declining-balance depreciation for the first, second, and third years. * ***Round to two places.***

Remember to multiply the depreciation rate by two.

Use multioperations and the % minus feature on your calculator.

	Purchase Price	Life	Rate of Depreciation	Depreciation 1st Year	Depreciation 2nd Year	Depreciation 3rd Year
1.	$4,200	6	16.67%	________	________	________
2.	$730	10	10%	________	________	________
3.	$8,950	8	12.5%	________	________	________

	Purchase Price	Life	Rate of Depreciation	Depreciation 1st Year	2nd Year	3rd Year
4.	$16,480	12	8.33%	________	________	________
5.	$250,000	25	4%	________	________	________

In the following problems, find the declining-balance depreciation and book value of each asset for the year indicated.

Remember, do not depreciate the asset below salvage value.

	Purchase Price	Rate of Depreciation	Salvage Value	Year of Life	Depreciation	Book Value
6.	$18,250	20%	$1,925	5	________	________
7.	$3,500	25%	$250	4	________	________

	Purchase Price	Rate of Depreciation	Salvage Value	Year of Life	Depreciation	Book Value
8.	$8,395	10%	$750	3	________	________
9.	$11,615	20%	$2,460	4	________	________
10.	$5,000	16.67%	$230	2	________	________

Sum-of-the-Years-Digits Method

When a business uses *sum-of-the-years-digits depreciation* method, the asset will depreciate more during the first years of useful life just as in declining-balance depreciation. The total amount that is depreciated is the same as in straight-line depreciation, but book value decreases faster in the first year than in the last. The depreciable amount (purchase price − salvage value) is used in determining the yearly depreciation.

The rate of depreciation is stated as a fraction. The denominator of the fraction is the sum of the years of useful life.

Example: An asset with five years estimated life will have a depreciation rate whose denominator is 15.

$5 + 4 + 3 + 2 + 1 = 15$ (denominator)

The denominator of the rate can also be found by using the formula:

$$\frac{n(n + 1)}{2} \quad (n = \text{the number of years of life of the asset})$$

Example: An asset with five years estimated life will have a depreciation rate whose denominator is 15.

$$\frac{n(n + 1)}{2} = \frac{5(5 + 1)}{2} = \frac{5(6)}{2} = \frac{30}{2} = 15 \quad \text{(denominator)}$$

The rate will change each year with the numerator indicating the years of life left at the beginning of each year.

First year rate $= \frac{5}{15} = \frac{1}{3}$
Second year rate $= \frac{4}{15}$
Third year rate $= \frac{3}{15} = \frac{1}{5}$
Fourth year rate $= \frac{2}{15}$
Fifth year rate $= \frac{1}{15}$

To determine the depreciation, multiply the depreciable amount times each year's rate. The first year's rate of $\frac{5}{15}$ will give more depreciation than the fifth year's rate of $\frac{1}{15}$.

Example: A $13,000 asset with a $3,000 salvage value has a five-year life. The depreciation is computed as follows:

Depreciable Amount: $13,000 – $3,000 = $10,000

First year depreciation: $10,000 × $\frac{5}{15}$ = $3,333.33
Second year depreciation: $10,000 × $\frac{4}{15}$ = $2,666.67
Third year depreciation: $10,000 × $\frac{3}{15}$ = $2,000.00
Fourth year depreciation: $10,000 × $\frac{2}{15}$ = $1,333.33
Fifth year depreciation: $10,000 × $\frac{1}{15}$ = $666.67
$10,000.00

It is easy to find any given year's depreciation using this method. In the above example, if you need to know the third year deprecation amount, simply multiply the depreciable amount by $\frac{3}{15}$. To find the last year's depreciation without calculating the other years, multiply by $\frac{1}{15}$.

Find the depreciation for the year indicated using sum-of-the-years-digits depreciation.

	Original Cost	Estimated Life	Salvage Value	Year	Depreciation
1.	$4,500	3 years	$900	2	________
2.	$14,300	7 years	$1,200	5	________
3.	$4,320	4 years	-0-	4	________

For the next two problems, find the denominator of the rate using the formula: $\frac{n(n + 1)}{2}$

	Original Cost	Estimated Life	Salvage Value	Year	Depreciation
4.	$45,000	10 years	$5,000	8	________
5.	$7,800	4 years	$800	1	________

To find accumulated depreciation, add the fractions for each year involved to form an **accumulated depreciation rate.**

Example: A $13,000 asset with a $3,000 salvage value has a five-year life. To find the accumulated depreciation rate at the end of three years, add the fractions for the first three years.

$\frac{5}{15} + \frac{4}{15} + \frac{3}{15} = \frac{12}{15}$ (accumulated depreciation rate)

Use the accumulated depreciation rate to compute the accumulated depreciation.

$\$10{,}000 \times \frac{12}{15} = \$8{,}000$ (accumulated depreciation at the end of three years)

Find the accumulated depreciation rate and accumulated depreciation for the year of life indicated. Reduce fractions to lowest terms. Use the formula to find the denominator.

	Original Cost	Estimated Life	Year of Life	Accumulated Depreciation Rate	Accumulated Depreciation
6.	$2,000	4	3	________	________
7.	$9,000	8	2	________	________

	Original Cost	Estimated Life	Year of Life	Accumulated Depreciation Rate	Accumulated Depreciation
8.	$8,800	11	5	______	______
9.	$12,500	15	4	______	______
10.	$16,800	7	2	______	______

Gain or Loss on Sale

If a company sells or trades an asset for more than the book value, the difference must be recorded as income to the business and is a taxable **gain on sale**. If a company sells or trades an asset for less than the book value, the difference is a **loss on sale** and can be recorded as an expense of doing business or a loss.

Example: A van purchased for $15,000 is estimated to have a useful life of five years with a $2,500 salvage value. The company uses the straight-line method of depreciation. At the end of four years accumulated depreciation is $10,000 and book value is $5,000. If the van is sold for $6,000 at the end of the fourth year, there is a $1,000 gain on the sale.

$6,000 (selling price) – $5,000 (book value) = $1,000 (gain)

If the company sells the van for $3,500, then there is a $1,500 loss.

$5,000 (book value) – $3,500 (selling price) = $1,500 (loss)

Find the gain or loss on sale or trade-in for each of the following assets. Round answers to two decimal places.

	Cost	Accumulated Depreciation	Sale/Trade-in Price	Gain	Loss
1.	$20,000	$17,250	$3,225	______	______
2.	$7,095	$6,000	$750	______	______

	Cost	Accumulated Depreciation	Sale/Trade-in Price	Gain	Loss
3.	$13,000	$11,195	$805	______	______
4.	$4,550	$4,000	$750	______	______
5.	$8,135	$6,925	$2,000	______	______

EVALUATING YOUR SKILLS

The following problems evaluate your skills in the material covered in Units 1 through 14. Some of the problems will evaluate your ability to work problems mentally while others will evaluate your ability to use your calculator. Work the problems as quickly as you can.

1. Arrange these numbers in order so that the largest number is first and the smallest last.

.8 .073 .29 .6 ______

2. Change .6 to a common fraction. ______

3. Write 2.1 as a percent. ______

4. $\frac{7}{12} = \frac{?}{24}$ ______

5. June sales were $19,450. July sales were $21,320. What was the percent increase? ______

6. Add $9\frac{1}{2} + 4\frac{5}{6} + 2\frac{1}{12} + 16\frac{1}{4}$. ______

7. Multiply $16,000 × .0125. ______

8. Use the cancellation method to divide $13\frac{1}{2}$ by $\frac{3}{8}$. ______

9. A salary of $16,900 a year is equal to how much a week? __________

10. Estimate the product of 89.2 × .4. __________

11. Estimate the quotient of 7,980 ÷ 240. __________

12. Find the cash price if paid within the discount period: list price is $1,330; trade discount is 10, 15; terms are 2/30, n/60. __________

13. Watson's bank statement balance on May 1 was $847.29. The checkbook balance was $916.75. A check for $32.10 has not been recorded in the check register. There was an outstanding deposit of $103.18. A service charge of $5 was shown on the bank statement. Outstanding checks were: $38.82, $14.90, $17.10. What is Watson's reconciled balance? __________

BANK RECONCILIATION
May 1, 19--

14. Beth Sanders earns a monthly salary of $720, a 4% commission on all sales plus an additional 2% on all sales over $12,000. Her sales for last month were $20,200. What were her total earnings last month? __________

15. A coat that costs $113 has a markup on selling price of 21%. What is the selling price? __________

APPLYING YOUR KNOWLEDGE

Using the information given, calculate the correct answers.

* *Set decimal selector on 2.*

1. An asset costing $7,000 has an estimated life of four years and a salvage value of $600. If depreciation is calculated using the straight-line method, what is the annual depreciation?

   ____

2. An asset costing $13,095 has an estimated life of six years and a salvage value of $1,250. What is the depreciation for the third year using the sum-of-the-years-digits method?

   ____

3. An asset costing $17,000 has an estimated salvage value of $3,125 and an estimated life of four years. What is the first year depreciation using the declining-balance method?

   ____

4. An asset costing $9,250 has a salvage value of $1,900 and an estimated life of three years. What is the depreciation for each of the three years using the sum-of-the-years-digits method?
   a. First year

   ____

   b. Second year

   ____

   c. Third year

   ____

5. The Heat Company uses the declining-balance method when calculating depreciation and a rate based on straight-line depreciation. They purchased five word processors at a total cost of $45,000. The total estimated salvage value is $5,000, and the estimated life is five years.
   a. What is total accumulated depreciation at the end of four years?

   ____

   b. What is the book value at the end of five years?

   ____

   c. What is the amount of depreciation recorded for the fifth year?

   ____

6. The Zong Corporation purchased equipment costing $56,296. Its useful life is estimated to be eight years with a salvage value of $4,700. What is third year depreciation using:
   a. Straight-line?

   ____

   b. Declining-balance?

   ____

   c. Which method yields the larger depreciation the third year?

   ____

7. An asset costing $14,970 has an estimated useful life of six years and a salvage value of $2,380. Using the sum-of-the-years-digits method determine the following:
   a. Fourth year depreciation

   ____

   b. Accumulated depreciation at end of fifth year (Hint: Find sixth year depreciation and subtract from beginning book value.)

   ____

   c. Book value at end of sixth year

   ____

8. An asset costing $11,585 has an estimated useful life of eight years and a salvage value of $500. The asset is sold for $5,193 at the end of six years. Use the sum-of-the-years-digits method.
   a. Is there a gain or loss on the sale?

   ____

   b. What is the amount of the gain or loss?

   ____

9. A copier costing $4,300 has a useful life of four years and a trade-in value of $200. It is traded in after $3\frac{1}{2}$ years for $300. Use straight-line depreciation.
   a. What is the accumulated depreciation at the end of $3\frac{1}{2}$ years?

   ____

   b. What is the book value at the end of $3\frac{1}{2}$ years?

   ____

   c. Is there a gain or a loss?

   ____

   d. What is the amount of the gain or loss?

   ____

ADVANCED APPLICATIONS

Using the information given, calculate the correct answers.

1. The Fourte Company purchased a computer for $25,000. The company plans to trade in the computer at the end of six years for $4,600. Determine accumulated depreciation at the end of four years using the following methods:
   a. Straight-line

   b. Sum-of-the-years-digits

   c. Declining-balance

   d. If the asset is sold at the end of four years for $13,000, which method will yield the largest gain?

   e. What is the amount of the gain?

2. A copier has an estimated useful life of four years and a salvage value of $300. Depreciation for the second year is $1,260. Determine the cost of the copier if the sum-of-the-years-digit method is used to compute depreciation.

3. A company prepares a record of the depreciation that will be recorded each year for each asset using the declining-balance method. This record is called a depreciation schedule. Prepare a depreciation schedule for equipment costing $27,000 with an estimated useful life of eight years and an estimated salvage value of $3,225.

Year	Rate of Depreciation	Amount of Depreciation	Book Value	Accumulated Depreciation
1				
2				
3				
4				
5				
6				
7				
8				

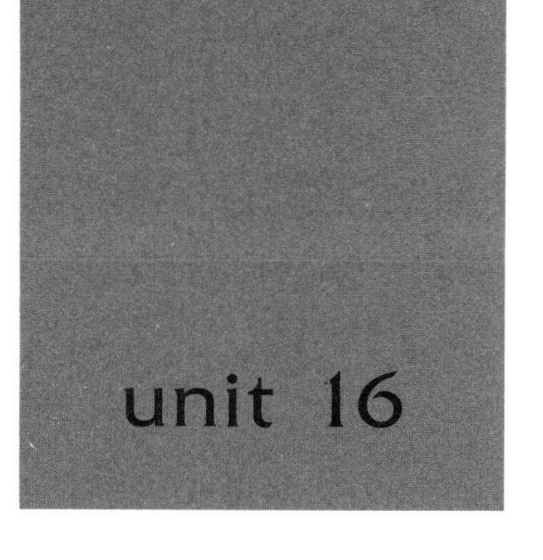

unit 16 Credit

"In order to make your dreams come true you must be wide awake."

The use of credit in our society has allowed businesses to expand and people to enjoy a higher standard of living. The decision to purchase an item at a later date or to buy it now using credit must be made carefully. The wise and proper use of credit is important if individuals and businesses are to survive in our complex economy. This means understanding credit terms, knowing the cost of credit, and being careful not to overextend. It also means shopping around for the best credit terms.

After studying this unit, you should be able to:

1. compute simple interest
2. calculate the amount financed on an installment loan
3. use a chart to calculate monthly payments on an installment loan
4. calculate the rate when the finance charge is known
5. use a formula or a table to calculate the annual percentage rate
6. calculate the finance charge on a revolving charge account
7. determine the down payment, calculate the monthly payments using a table, and find the actual cost of a mortgage loan

MATH TERMS AND CONCEPTS

The following terminology is used in this unit. Become familiar with the meaning of each term so that you understand its usage as you develop math skills.

1. **Amount Financed**—amount of the principal minus the down payment.
2. **Annual**—occurring once a year.
3. **Annual Percentage Rate**—the cost of credit on a yearly basis.
4. **Down Payment**—amount applied to the cash price at the time of purchase on an installment plan.
5. **Finance Charge**—a charge for borrowing or paying over a period of time.
6. **Installment Loan**—a loan on which equal periodic payments are made.
7. **Installment Price**—principal plus interest and all fees.

8. **Interest**—the cost of financing a debt over a period of time.
9. **Mortgage**—a loan on real property where the lender has the right to receive ownership if loan payments are not made.
10. **Principal**—the amount borrowed.
11. **Proceeds**—principal of a loan minus interest and fees.
12. **Revolving Charge Account**—a type of credit that allows the purchase of products or services with the agreement that the balance will be paid in full or that a finance charge will be paid on the unpaid balance.
13. **Simple Interest**—interest calculated on the principal only.

Simple Interest

Using credit costs money. The **principal** plus **interest** is repaid for the time the money is used. The formula to compute **simple interest** has four elements:

$$\text{Interest} = \text{Principal} \times \text{Rate} \times \text{Time}$$
$$I = P \times R \times T$$

This formula is similar to the PBR formulas you learned in Unit 8. The interest is the percentage, the principal is the base, and the rate is the percent. The fourth element is the length of time of the loan. The time is based on one year.

Example: $250 is borrowed for one year at $12\frac{1}{2}\%$. The interest is calculated as:

$$P \times R \times T = I$$
$$\$250 \times .125 \times 1 = \$31.25$$

The amount to be repaid is: $\$250 + \$31.25 = \$281.25$

Find the amount of interest and amount due for each of the following.

	Amount Borrowed	Rate	Time	Amount of Interest	Amount Due
1.	$2,000	16%	3 years	________	________
2.	$750	$18\frac{1}{2}\%$	1 year	________	________
3.	$1,230	$17\frac{3}{4}\%$	2 years	________	________
4.	$400	12.5%	$1\frac{1}{2}$ years	________	________
5.	$925	15%	2 years	________	________

Many businesses borrow money for a specified number of days. The days are expressed as a fractional part of a year. Most lending institutions use the bankers' year which assumes 360 days in a year or 30 days in each month. Some use the exact days of 365 days in a year. The bankers' year will be used in this unit.

The time (T) is found by forming a fraction using 360 as the denominator and the number of days of the loan as the numerator. Use this formula to find the interest for a given number of days:

$$\text{Interest} = \text{Principal} \times \text{Rate} \times \frac{\text{Number of Days}}{360}$$

Example: The amount of interest on a $350 loan borrowed for 23 days at 18% is:

$$P \times R \times T = I$$
$$\$350 \times .18 \times \frac{23}{360} = \$4.03$$

Find the interest on the following loans.

Use multifactor multiplication followed by division.

When using the formula without the aid of the calculator, use the cancellation method of multiplication.

	Principal	Annual Interest Rate	Time	Interest
6.	$4,230	17%	50 days	____________
7.	$1,962	$16\frac{1}{4}$%	230 days	____________
8.	$12,860	$15\frac{3}{4}$%	128 days	____________
9.	$10,753	18%	95 days	____________
10.	$845	$17\frac{3}{4}$%	46 days	____________

When money is borrowed for a specific number of months, the months are expressed as a fractional part of a year. The time is found by forming a fraction using 12 as the denominator and the number of months of the loan as the numerator. Use this formula to find the interest for a given number of months:

$$\text{Interest} = \text{Principal} \times \text{Rate} \times \frac{\text{Number of Months}}{12}$$

Example: The interest on a $1,200 loan borrowed for 3 months at $16\frac{1}{2}$% is:

$$P \times R \times T = I$$
$$\$1{,}200 \times .165 \times \frac{3}{12} = \$49.50$$

* *Reduce the fraction to lowest terms: $\frac{1}{4}$.*

Find the interest on the following loans.

Loans for 30, 60, 90 days are treated as loans for 1, 2, or 3 months.

	Principal	Annual Interest Rate	Time	Interest
11.	$2,790	$14\frac{3}{4}$%	4 months	________
12.	$8,342	15%	60 days	________
13.	$5,600	$17\frac{1}{4}$%	6 months	________
14.	$1,130	19%	4 months	________
15.	$800	$12\frac{3}{4}$%	1 month	________

Installment Loans

When an item is purchased on credit, it is usually paid for in equal monthly installments. A loan may also be repaid on the installment plan. A **down payment** is usually required. The interest is calculated on the **amount financed**, or the amount of principal minus the down payment.

Selling Price − Down Payment = Amount Financed

Example: Calculate the amount financed on a $1,500 purchase that requires a 15% down payment.

Solve in one step by using the %− feature.

a. Calculate the down payment: $1,500 × .15 = $225
b. Subtract the down payment from the selling price to get the amount financed: $1,500 − $225 = $1,275

Tables similar to the one in Illustration 16-1 are used in finding the monthly payment for an **installment loan**. Using the **annual percentage rate** and length of the loan, find the decimal for the monthly payment per $1 borrowed. Then multiply the decimal by the amount borrowed to determine the monthly payment.

Monthly payment per $1 borrowed

	6 Mo.	12 Mo.	18 Mo.	24 Mo.
10%	.17156	.08792	.06006	.04614
12%	.17255	.08885	.06098	.04707
14%	.17354	.08979	.06192	.04801
16%	.17453	.09073	.06286	.04896
18%	.17554	.09168	.06381	.04992
20%	.17652	.09263	.06476	.05090
22%	.17752	.09359	.06573	.05188

Illustration 16-1. Monthly Payment Per $1 Borrowed.

Example: Calculate the monthly payment on $2,220 borrowed for 24 months at 16%:

a. Find the monthly payment per $1 borrowed: .04896
b. Multiply the decimal by the amount borrowed to find the monthly payment on $2,220: .04896 × $2,220 = $108.69

Use Illustration 16-1 to find the monthly payment for each of the following loans.

	Principal	Rate	Time	Monthly Payment
1.	$600	18%	12 months	______
2.	$1,300	20%	18 months	______
3.	$865	16%	24 months	______
4.	$2,476	22%	24 months	______

Although each monthly payment is the same, the portion applied to the interest gradually decreases while the portion applied to the principal increases. If you need to know the principal balance of a loan at any given time:

1. Use the simple interest formula to calculate the amount of interest for the first month.
2. Find the amount applied to the principal. To do this, subtract the monthly interest from the monthly payment.
3. Find the new balance. To do this, subtract the amount applied to the principal from the balance.
4. Using the *new* balance, calculate interest for the second month and repeat Steps 2 and 3.
5. Continue repeating Step 4 to find the principal balance for each of the following months.

The total of the interest payments at any time subtracted from the total payments at that time should equal the beginning principal balance.

Example: Monthly payments for a $1,000 loan to be repaid in 12 equal payments at 16% interest are $90.73. To find the principal balance at the end of the first month:

1. Calculate the interest for the first month.
   $\$1,000 \times .16 \times \frac{1}{12} = \$13.33$
2. Find the amount applied to the principal.
   $90.73 − $13.33 = $77.40
3. Find the balance owed after the first payment.
   $1,000 − $77.40 = $922.60
4. To find the principal balance after the second payment, calculate the interest on the new balance for one month.
   $\$922.60 \times .16 \times \frac{1}{12} = \$12.30$
5. Repeat Steps 2 and 3 to find the principal balance after the second payment.
   $90.73 − $12.30 = $78.43
   $922.60 − $78.43 = $844.17

Complete the following chart to find the new principal balance after each payment for a six-month loan at 20% interest for $500. Use the monthly payment table in Illustration 16-1 on page 284.

	End of Month	Principal Balance	Interest	Payment on Principal	New Balance
1.	One	$500	________	________	________
2.	Two	________	________	________	________
3.	Three	________	________	________	________
4.	Four	________	________	________	________
5.	Five	________	________	________	________
6.	Six	________	________	________	________

* *Any difference is paid in the final payment. The last payment is $88.27.*

Notice the total interest paid subtracted from the total payments equals the beginning principal balance of $500.

Finance Charge

The sum of interest and any other fees charged to the customer on an installment loan is called the **finance charge**. To find the finance charge, first multiply the monthly payments by the number of payments and add any down payment to obtain the **installment price**. Then subtract the cash price from the installment price.

Monthly Payments × Number of Payments = Total Amount to be Repaid

Total Amount Repaid + Down Payment = Installment Price

Installment Price − Cash Price = Finance Charge

Example: The monthly payment on a 24-month loan for $7,000 with a down payment of 20% is $279.55. The finance charge is $1,109.20.

Total Amount to be Repaid: $279.55 × 24 = $6,709.20
Installment Price: $6,709.20 + $1,400 = $8,109.20
Finance Charge: $8,109.20 − $7,000 = $1,109.20

Find the amount to be repaid and the finance charge.

	Cash Price	Down Payment	Amount Financed	Monthly Payment	Number of Payments	Amount to be Repaid	Finance Charge
1.	$600	$75	$525	$91.11	6	________	________
2.	$3,000	$200	$2,800	$181.86	18	________	________
3.	$6,150	$500	$5,650	$276.62	24	________	________
4.	$1,100	$225	$875	$81.89	12	________	________
5.	$750	$50	$700	$35.63	24	________	________

This is a variation of the simple interest formula: $I = P \times R \times T$.

To find the interest rate when the finance charge is known, use the formula: Rate = Interest ÷ (Principal × Time)

$$R = \frac{I}{P \times T}$$

In the example above the amount to be financed on a 24-month, $7,000 loan with a $1,400 down payment is $5,600. The finance charge is $1,109.20. The rate is 9.9%.

$$R = \frac{\$1{,}109.20}{\$5{,}600 \times 2} = \frac{\$1{,}109.20}{\$11{,}200} = 9.9\%$$

Find the rate. Round to the nearest hundredth of a percent.

	Finance Charge	Amount Financed	Time	Rate
6.	$21.66	$525	6 months	________
7.	$423.48	$2,850	18 months	________
8.	$988.88	$5,650	2 years	________

	Finance Charge	Amount Financed	Time	Rate
9.	$107.68	$875	1 year	________
10.	$155.12	$700	2 years	________

Annual Percentage Rate

On installment loans, the borrower is charged a finance charge on the full amount of the loan for the length of the entire payment period and not on the balance owed as payments are made. The Truth-in-Lending Act of 1969 requires the lender to disclose the finance charge to the borrower, which must be expressed as the **annual percentage rate** (APR). Tables, such as in Illustration 16-2 on page 289, are used in business to let consumers know exactly what they must pay for the use of credit.

Example: The finance charge for a $5,600 loan is $1,109.20. The loan is paid monthly for 24 months. Using Illustration 16-2 on page 289, find the rate.

1. Divide the finance charge by the amount financed and multiply by 100: $1,109.20 ÷ $5,600 × 100 = 19.81
2. Read down the number of payments column to 24.
3. Read across the table to the number closest to 19.81: Do you find the 19.82?
4. Read up the column to find the rate at the top: The annual percentage rate is 18%.

Using Illustration 16-2, find the annual percentage rate (APR) for each of the following.

	Finance Charge	Amount Financed	Number of Payments	APR
1.	$21.66	$525	6	________
2.	$473.48	$2,850	22	________
3.	$988.88	$5,650	24	________
4.	$107.68	$875	15	________
5.	$105.00	$700	18	________

Revolving Charge Accounts

Most oil companies, banks, and department stores use **revolving charge accounts** for credit card customers. A revolving charge account allows a customer to pay the whole bill every month and avoid any finance charge. A customer who cannot pay the entire amount due is required to make a minimum payment and pay the balance over a period of time.

NUMBER OF PAYMENTS	ANNUAL PERCENTAGE RATE																				
	10.00%	10.50%	11.00%	11.50%	12.00%	12.50%	13.00%	13.50%	14.00%	14.50%	15.00%	15.50%	16.00%	16.50%	17.00%	17.50%	18.00%	18.50%	19.00%	19.50%	20.00%
	(FINANCE CHARGE PER $100 OF AMOUNT FINANCED)																				
1	0.83	0.87	0.92	0.96	1.00	1.04	1.08	1.12	1.17	1.21	1.25	1.29	1.33	1.37	1.42	1.46	1.50	1.54	1.58	1.62	1.67
2	1.25	1.31	1.38	1.44	1.50	1.57	1.63	1.69	1.75	1.82	1.88	1.94	2.00	2.07	2.13	2.19	2.26	2.32	2.38	2.44	2.51
3	1.67	1.76	1.84	1.92	2.01	2.09	2.17	2.26	2.34	2.43	2.51	2.59	2.68	2.76	2.85	2.93	3.01	3.10	3.18	3.27	3.35
4	2.09	2.20	2.30	2.41	2.51	2.62	2.72	2.83	2.93	3.04	3.14	3.25	3.36	3.46	3.57	3.67	3.78	3.88	3.99	4.10	4.20
5	2.51	2.64	2.77	2.89	30.2	3.15	3.27	3.40	3.53	3.65	3.78	3.91	4.04	4.16	4.29	4.42	4.54	4.67	4.80	4.93	5.06
6	2.49	3.08	3.23	3.38	3.53	3.68	3.83	3.97	4.12	4.27	4.42	4.57	4.72	4.87	5.02	5.17	5.32	5.46	5.61	5.76	5.91
7	3.36	3.53	3.70	3.87	4.04	4.21	4.38	4.55	4.72	4.89	5.06	5.23	5.40	5.58	5.75	5.92	6.09	6.26	6.43	6.60	6.78
8	3.79	3.98	4.17	4.36	4.55	4.74	4.94	5.13	5.32	5.51	5.71	5.90	6.09	6.29	6.48	6.67	6.87	7.06	7.26	7.45	7.64
9	4.21	4.43	4.64	4.85	5.07	5.28	5.49	5.71	5.92	6.14	6.35	6.57	6.78	7.00	7.22	7.43	7.65	7.87	8.08	8.30	8.52
10	4.64	4.88	5.11	5.35	5.58	5.82	6.05	6.29	6.53	6.77	7.00	7.24	7.48	7.72	7.96	8.19	8.43	8.67	8.91	9.15	9.39
11	5.07	5.33	5.58	5.84	6.10	6.36	6.62	6.88	7.14	7.40	7.66	7.92	8.18	8.44	8.70	8.96	9.22	9.49	9.75	10.01	10.28
12	5.50	5.78	6.06	6.34	6.62	6.90	7.18	7.46	7.74	8.03	8.31	8.59	8.88	9.16	9.45	9.73	10.02	10.30	10.59	10.87	11.16
13	5.93	6.23	6.53	6.84	7.14	7.44	7.75	8.05	8.36	8.66	8.97	9.27	9.58	9.89	10.20	10.50	10.81	11.12	11.43	11.74	12.05
14	6.36	6.69	7.01	7.34	7.66	7.99	8.31	8.64	8.97	9.30	9.63	9.96	10.29	10.62	10.95	11.28	11.61	11.95	12.28	12.61	12.95
15	6.80	7.14	7.49	7.84	8.19	8.53	8.88	9.23	9.59	9.94	10.29	10.64	11.00	11.35	11.71	12.06	12.42	12.77	13.13	13.49	13.85
16	7.23	7.60	7.97	8.34	8.71	9.08	9.46	9.83	10.20	10.58	10.95	11.33	11.71	12.09	12.46	12.84	13.22	13.60	13.99	14.37	14.75
17	7.67	8.06	8.45	8.84	9.24	9.63	10.03	10.43	10.82	11.22	11.62	12.02	12.42	12.83	13.23	13.63	14.04	14.44	14.85	15.25	15.66
18	8.10	8.52	8.93	9.35	9.77	10.19	10.61	11.03	11.45	11.87	12.29	12.72	13.14	13.57	13.99	14.42	14.85	15.28	15.71	16.14	16.57
19	8.54	8.98	9.42	9.86	10.30	10.74	11.18	11.63	12.07	12.52	12.97	13.41	13.86	14.31	14.76	15.22	15.67	16.12	16.58	17.03	17.49
20	8.98	9.44	9.90	10.37	10.83	11.30	11.76	12.23	12.70	13.17	13.64	14.11	14.59	15.06	15.54	16.01	16.49	16.97	17.45	17.93	18.41
21	9.42	9.90	10.39	10.88	11.36	11.85	12.34	12.84	13.33	13.82	14.32	14.82	15.31	15.81	16.31	16.81	17.32	17.82	18.33	18.83	19.34
22	9.86	10.37	10.88	11.39	11.90	12.41	12.93	13.44	13.96	14.48	15.00	15.52	16.04	16.57	17.09	17.62	18.15	18.68	19.21	19.74	20.27
23	10.30	10.84	11.37	11.90	12.44	12.97	13.51	14.05	14.59	15.14	15.68	16.23	16.78	17.32	17.88	18.43	18.98	19.54	20.09	20.65	21.21
24	10.75	11.30	11.86	12.42	12.98	13.54	14.10	14.66	15.23	15.80	16.37	16.94	17.51	18.09	18.66	19.24	19.82	20.40	20.98	21.56	22.15
25	11.19	11.77	12.35	12.93	13.52	14.10	14.69	15.28	15.87	16.46	17.06	17.65	18.25	18.85	19.45	20.05	20.66	21.27	21.87	22.48	23.10
26	11.64	12.24	12.85	13.45	14.06	14.67	15.28	15.89	16.51	17.13	17.75	18.37	18.99	19.62	20.24	20.87	21.50	22.14	22.77	23.41	24.04
27	12.09	12.71	13.34	13.97	14.60	15.24	15.87	16.51	17.15	17.80	18.44	19.09	19.74	20.39	21.04	21.69	22.35	23.01	23.67	24.33	25.00
28	12.53	13.18	13.84	14.49	15.15	15.81	16.47	17.13	17.80	18.47	19.14	19.81	20.48	21.16	21.84	22.52	23.20	23.89	24.58	25.27	25.96
29	12.98	13.66	14.33	15.01	15.70	16.38	17.07	17.75	18.45	19.14	19.83	20.53	21.23	21.94	22.64	23.35	24.06	24.77	25.49	26.20	26.92
30	13.43	14.13	14.83	15.54	16.24	16.95	17.66	18.38	19.10	19.81	20.54	21.26	21.99	22.72	23.45	24.18	24.92	25.66	26.40	27.14	27.89
31	13.89	14.61	15.33	16.06	16.79	17.53	18.27	19.00	19.75	20.49	21.24	21.99	22.74	23.50	24.26	25.02	25.78	26.55	27.32	28.09	28.86
32	14.34	15.09	15.84	16.59	17.35	18.11	18.87	19.63	20.40	21.17	21.95	22.72	23.50	24.28	25.07	25.86	26.65	27.44	28.24	29.04	29.84
33	14.79	15.57	16.34	17.12	17.90	18.69	19.47	20.26	21.06	21.85	22.65	23.46	24.26	25.07	25.88	26.70	27.52	28.34	29.16	29.99	30.82
34	15.25	16.05	16.85	17.65	18.46	19.27	20.08	20.90	21.72	22.54	23.37	24.19	25.03	25.86	26.70	27.54	28.39	29.24	30.09	30.95	31.80
35	15.70	16.53	17.35	18.18	19.01	19.85	20.69	21.53	22.38	23.23	24.08	24.94	25.79	26.66	27.52	28.39	29.27	30.14	31.02	31.91	32.79
36	16.16	17.01	17.86	18.71	19.57	20.43	21.30	22.17	23.04	23.92	24.80	25.68	26.57	27.46	28.35	29.25	30.15	31.05	31.96	32.87	33.79
37	16.62	17.49	18.37	19.25	20.13	21.02	21.91	22.81	23.70	24.61	25.51	26.42	27.34	28.26	29.18	30.10	31.03	31.97	32.90	33.84	34.79
38	17.08	17.98	18.88	19.78	20.69	21.61	22.52	23.45	24.37	25.30	26.24	27.17	28.11	29.06	30.01	30.96	31.92	32.88	33.85	34.82	35.79
39	17.54	18.46	19.39	20.32	21.26	22.20	23.14	24.09	25.04	26.00	26.96	27.92	28.89	29.87	30.85	31.83	32.81	33.80	34.80	35.80	36.80
40	18.00	18.95	19.90	20.86	21.82	22.79	23.76	24.73	25.71	26.70	27.69	28.68	29.58	30.69	31.68	32.69	33.71	34.73	35.75	36.78	37.81
41	18.47	19.44	20.42	21.40	22.39	23.38	24.38	25.38	26.39	27.40	28.41	29.44	30.46	31.49	32.52	33.55	34.61	35.66	36.71	37.77	38.83
42	18.93	19.93	20.93	21.94	22.96	23.98	25.00	26.03	27.06	28.10	29.15	30.19	31.25	32.31	33.37	34.44	35.51	36.59	37.67	38.76	39.85
43	19.40	20.42	21.45	22.49	23.53	24.57	25.62	26.68	27.74	28.81	29.88	30.96	32.04	33.13	34.22	35.31	36.42	37.52	38.63	39.75	40.87
44	19.86	20.91	21.97	23.03	24.10	25.17	26.25	27.33	28.42	29.52	30.62	31.72	32.83	33.95	35.07	36.19	37.33	38.46	39.60	40.75	41.90
45	20.33	21.41	22.49	23.58	24.67	25.77	26.88	27.99	29.11	30.23	31.36	32.49	33.63	34.77	35.92	37.08	38.24	39.41	40.58	41.75	42.94
46	20.80	21.90	23.01	24.13	25.25	26.37	27.51	28.65	29.79	30.94	32.10	33.26	34.43	35.60	36.78	37.96	39.16	40.35	41.55	42.76	43.98
47	21.27	22.40	23.53	24.68	25.82	26.98	28.14	29.31	30.48	31.66	32.84	34.03	35.23	36.43	37.64	38.86	40.08	41.30	42.54	43.77	45.02
48	21.74	22.90	24.06	25.23	26.40	27.58	28.77	29.97	31.17	32.37	33.59	34.81	36.03	37.27	38.50	39.75	41.00	42.26	43.52	44.79	46.07
49	22.21	23.39	24.58	25.78	26.98	28.19	29.41	30.63	31.86	33.09	34.34	35.59	36.84	39.10	39.37	40.65	41.93	43.22	44.51	45.81	47.12
50	22.69	23.89	25.11	26.33	27.56	28.80	30.04	31.29	32.55	33.82	35.09	36.37	37.65	38.94	40.24	41.55	42.86	44.18	45.50	46.83	48.17

Illustration 16-2. Annual Percentage Rate Table (APR).

A credit limit—a maximum amount that can be bought on credit—is established for each individual charge account based on the individual's salary and credit references. Charge purchases may be made up to the limit. If the balance is paid within a certain period of time—often 30 days—there is usually no finance charge. A finance charge is calculated on any unpaid balance and is usually stated as both a monthly rate, such as $1\frac{1}{2}$ percent per month, and as an annual percentage rate (APR), such as 18 percent annually.

The rates may vary depending on the balance due. A chart, such as the one in Illustration 16-3, is usually given on the billing statement to show the rates charged.

2.16% per month or 26% annual for up to $480
1.75% per month or 21% annual for $480 to $1,600
1.25% per month or 15% annual for excess of $1,600

Illustration 16-3. Finance Charge Table.

Example: Sue Lund's unpaid balance is $423. Using the table in Illustration 16-3, the finance charge is $9.14 ($423 × 2.16%). The account balance would be $423 + $9.14 = $432.14.

Use the interest rate table in Illustration 16-3 to compute the finance charge on each unpaid balance.

	Unpaid Balance	Finance Charge	New Balance
1.	$373	______	______
2.	$1,227	______	______
3.	$872	______	______
4.	$320	______	______
5.	$514	______	______

Mortgage Loans

A mortgage loan for the purchase of a home or building is usually financed from 10 to 30 years and is repaid in equal monthly payments. The loan requires a percent of the selling price as a down payment. If the borrower fails to make the monthly payments, the lender has the right to sell the property. To determine the price you can afford to pay for a home, follow this guideline: Buy a home that costs less than $2\frac{1}{2}$ times your annual gross income.

Example: Sara and Jason are considering purchasing a home for $62,500. Their combined annual income is $30,000. The bank requires a 20% down payment. The balance will be financed for 30 years at $10\frac{1}{2}$%.

a. Price they can afford to pay: $30,000 × 2.5 = $75,000
b. The down payment required: $62,500 × 20% = $12,500
c. Amount to be financed: $62,500 − $12,500 = $50,000

Lenders use tables to find the monthly payment necessary to repay a **mortgage**. A monthly mortgage payment table shows the monthly payment for principal and interest only (see Illustration 16-4). The monthly payment can also include property taxes, homeowner's insurance, and mortgage insurance.

* *Note that the amounts listed in Illustration 16-4 are per $1,000.*

	(Principal & Interest per One Thousand Dollars)														
Number of Years	10%	$10\frac{1}{2}$%	11%	$11\frac{1}{2}$%	$11\frac{3}{4}$%	12%	$12\frac{1}{2}$%	$12\frac{3}{4}$%	13%	$13\frac{1}{2}$%	$13\frac{3}{4}$%	14%	$14\frac{1}{2}$%	$14\frac{3}{4}$%	15%
10...	13.22	13.50	13.78	14.06	14.21	14.35	14.64	14.79	14.94	15.23	15.38	15.53	15.83	15.99	16.14
12...	11.96	12.25	12.54	12.84	12.99	13.14	13.44	13.60	13.75	14.06	14.22	14.38	14.69	14.85	15.01
15...	10.75	11.06	11.37	11.69	11.85	12.01	12.33	12.49	12.66	12.99	13.15	13.32	13.66	13.83	14.00
17...	10.22	10.54	10.86	11.19	11.35	11.52	11.85	12.02	12.19	12.53	12.71	12.88	13.23	13.41	13.58
20...	9.66	9.99	10.33	10.67	10.84	11.02	11.37	11.54	11.72	12.08	12.26	12.44	12.80	12.99	13.17
22...	9.39	9.73	10.08	10.43	10.61	10.78	11.14	11.33	11.51	11.87	12.06	12.24	12.62	12.81	12.99
25...	9.09	9.45	9.81	10.17	10.35	10.54	10.91	11.10	11.28	11.66	11.85	12.04	12.43	12.62	12.81
30...	8.78	9.15	9.53	9.91	10.10	10.29	10.68	10.87	11.07	11.46	11.66	11.85	12.25	12.45	12.65
35...	8.60	8.99	9.37	9.77	9.96	10.16	10.56	10.76	10.96	11.36	11.56	11.76	12.17	12.37	12.57

Illustration 16-4. Monthly Mortgage Payment Table.

Example: Sara and Jason are financing $50,000 for 30 years at $10\frac{1}{2}$% interest. What is their monthly payment?

a. Using Illustration 16-4, find the length of the loan and the interest rate.
b. Multiply the amount given in the table times the number of thousands of the loan: $50 × 9.15 = $457.50

Sara and Jason would make a monthly payment of $457.50.

Use Illustration 16-4 to find the monthly payment for a $60,000 loan financed at 11% for each of the following number of years.

1. 10 years __________

2. 20 years __________

3. 30 years __________

When interest is added to the amount of the loan, the home buyer will pay back about two to three times the amount borrowed.

Example: Find the actual amount repaid on a $40,000 loan financed for 30 years at $10\frac{1}{2}\%$.

a. Use Illustration 16-4 to find the monthly payment: $9.15 \times \$40 = \$366$
b. Find the total annual cost: $12 \times \$366 = \$4,392$
c. Find the total amount repaid: $\$4,392 \times 30 = \$131,760$
d. Find the total interest: $\$131,760 - \$40,000 = \$91,760$

Using Illustration 16-4, find the monthly payment, the amount to be repaid, and the total interest on a $50,000 loan at each of the following rates and times.

	Interest Rate	Time	Monthly Payment	Amount Repaid	Interest
4.	10%	20 years	________	________	________
5.	12%	10 years	________	________	________
6.	14%	30 years	________	________	________
7.	$11\frac{3}{4}\%$	17 years	________	________	________
8.	$13\frac{3}{4}\%$	22 years	________	________	________

Many new mortgage plans are available. With some, the interest changes during the life of the loan. It is wise to consider all the various plans.

Each month's mortgage payment has an amount applied to the principal and an amount of interest. The simple interest formula is used to calculate the interest each month. Then subtract the interest from the monthly payment to get the amount applied to the principal.

Example: A $30,000 home is financed for 30 years at 11%. The monthly payment is $285.90. Follow these steps to find the interest and the amount applied to the principal.

a. Find the interest: $\$30{,}000 \times 11\% \times \frac{1}{12} = \$275$
b. Find the amount paid on principal: $\$285.90 - \$275 = \$10.90$
c. Find the new principal balance: $\$30{,}000 - \$10.90 = \$29{,}989.10$

To find the portions for the second month, use the new principal balance to calculate interest.

$\$30{,}000 - \$10.90 = \$29{,}989.10$ (new balance)
$\$29{,}989.10 \times 11\% \times \frac{1}{12} = \$274.90$ (interest)
$\$285.90 - \$274.90 = \$11.00$ (paid on principal)

Using the information above, find the interest and the amount applied to the principal for the third and fourth months.

	Interest	Principal
9. Third Month	______	______
10. Fourth Month	______	______

EVALUATING YOUR SKILLS

The following problems evaluate your skills in the material covered in Units 1 through 15. Some of the problems will evaluate your ability to work problems mentally while others will evaluate your ability to use your calculator. Work the problems as quickly as you can.

1. Write 105% as a decimal. ______

2. Change $\frac{1}{3}$ to a percent. ______

3. $246{,}780 \times 10 =$ ______

4. 30 is what % of 90? ______

5. 2.74 is 82% of what amount? ______

6. Add: $.2 + 2.0 + .002 + 20$. ______

7. Estimate the product of $473 \times .2$. ______

8. 5 pints is equal to how many liters? * *Use the conversion chart, Appendix B, page 339.* ______

9. $5\frac{1}{7} \times 5\frac{1}{4} =$ ______

10. Write 45% as a fraction in lowest terms. ______

11. In one year, the United States exports $33.7 billion and imports $46.5 billion to Canada. What percent more was imported than exported? ______

12. Store A sells an item for a list price of $1,232 and offers a 15, 10, 5 trade discount. Store B sells the same item listed at the same price but with a trade discount of 20, 5.
    a. Which store has the lowest price? ______

    b. By how much? ______

APPLYING YOUR KNOWLEDGE

Solve the following problems using the information given. Use the illustrations and formulas in this unit when needed.

1. Joyce and Allen Summers have a mortgage loan of $65,000. The interest rate is $10\frac{1}{2}$% with a monthly payment of $614.25. How much of the first monthly payment is applied to the interest? ______

2. Benson's Furniture Store advertised a chair for a cash price of $250. The chair could be purchased with a down payment of $25 and monthly payments of $14.36 for 18 months.
    a. What is the finance charge? ______

    b. What is the APR using Illustration 16-2? ______

3. Shannon Gentry purchased a sewing machine for $32.50 down and $43.10 a month for 12 months. She could have purchased the sewing machine for $475 cash. How much more did she pay on the installment plan than she would have by paying cash? ______

4. A retailer charges $1\frac{3}{4}$% interest per month on customers' unpaid balance. What is the annual interest rate? ______

5. Randy Schmidt has a revolving charge account at AXC Department Store. * *Finance charges are computed on beginning month's balance.* On June 30 he has a balance of $341. He made the following purchases: July 5, $72; July 29, $24; and August 11, $111.50. On July 31 Randy paid $50 on his account. On August 31 he paid one-tenth of his balance. Use Illustration 16-3 to compute the finance charges for July and August. What is his new balance after his payment on August 31?

____

6. Ly Law Firm borrowed $174,000 on September 1 to buy a building. The loan was financed for 17 years at $12\frac{3}{4}\%$ interest.

a. What is the monthly payment? (Use Illustration 16-4.)

____

b. How much interest will be paid by November 30 with a 20 percent down payment?

____

7. A rancher in Kansas wants to begin breeding ostriches. He purchased a pair from Africa for a total cost of $65,000. He paid 30% down and will make 36 monthly payments of $1,624.10. What is the finance charge?

____

8. Using the information in Problem 7, find the interest rate.

____

ADVANCED APPLICATIONS

Using the information given, calculate the correct answers. Use the illustrations and formulas in the unit when needed.

1. The Mounfort Company purchased merchandise on June 3 for $2,093. The invoice terms are 2/10, n/30. Mounfort cannot pay the invoice within the discount period. If the company borrows the amount needed to pay the invoice within the discount period, it will need the loan for only 17 days. The interest rate is 12%.
   a. What is the amount of the discount if paid in 10 days? __________
   b. How much would Mounfort Company have to borrow to pay the invoice within the discount period? __________
   c. How much interest would be paid on the loan? __________
   d. How much would Mounfort Company save by taking out the loan and paying the invoice within the discount period? __________

2. The Buckner Company wants to buy a building costing $95,150. A down payment of $15,150 is required. The monthly payment will be $1,092.60, which includes $127 for insurance and taxes and $80 for outside maintenance. What is the rate of interest on the loan if the loan will be financed for 30 years? (Hint: Determine the principal and interest payment per thousand.) __________

3. The Tonner Company is buying a computer at a cash price of $9,450 and a down payment of $4,035. The remainder will be financed for 24 months at 12% interest.
   a. What is the amount financed? __________
   b. What is the finance charge? __________
   c. What is the monthly payment? __________
   d. What is the installment price? __________
   e. When tables are not available to find the APR, the following formula can be used. Use the formula to find the APR.

$$\frac{2 \times \text{Number of Payments in One Year} \times \text{Finance Charge}}{\text{Principal} \times (\text{Number of Payments} + 1)}$$

____

unit 17 Ratio and Proportion

"Ignorance is the night of the mind. Wake Up!"

Comparison of numbers is important in both personal and business activities. The ability to compare numbers and make decisions based on the comparisons is necessary in the survival and growth of a company. The use of ratios and proportions gives clear relationships between numbers used in the decision-making processes.

After studying this unit, you should be able to:

1. **express a relationship as a ratio**
2. **understand and set up proportions**
3. **apply ratios and proportions to problem-solving situations**

MATH TERMS AND CONCEPTS

The following terminology is used in this unit. Become familiar with the meaning of each term so that you understand its usage as you develop math skills.

1. **Proportion**—a statement that two ratios are equal.
2. **Ratio**—a comparison of two numbers.

Ratio

A **ratio** shows a relationship between two numbers. A ratio, like a fraction, has two terms. The first term is the number that is being compared and is the numerator when the ratio is written as a fraction. The second term is the number to which the first number is being compared and is the denominator when the ratio is written as a fraction. A ratio can be expressed as two terms separated by the word *to* or a colon, as a fraction, as a decimal, or as a percent.

Example: There were 18 working days during the month of June. The ratio of working days to the days in the month is expressed as 18 to 30 or 18:30. The same comparison can be expressed as a:

fraction: $\frac{18}{30}$ or $\frac{3}{5}$ were working days.
decimal: .60 were working days.
percent: 60% were working days.

Each term has the same meaning as the others.

* *A ratio or fraction suggests division.*

Solve the following problems.

1. A factory hires 225 men and 100 women on its assembly line. Compare the number of women to the number of men as a

   a. Ratio: ______________ b. Fraction: ______________

   c. Decimal: ______________ d. Percent: ______________

   e. Express as a ratio the number of women to the total number of workers. * *The second term is the total number of workers (100 + 225).*

   ____

   f. Express as a ratio the number of men to the total number of workers.

   ____

2. Express each of the following as a ratio in lowest terms.

   a. 10:14 ________ b. 20:16 ________ c. 24:8 ________

   d. 8:12 ________ e. 7:28 ________ f. 30:10 ________

When numbers are expressed as a ratio, they must be in the same units.

Example: When comparing 8 inches to 2 feet, the feet must first be changed to inches. The ratio is 8 to 24, 8:24, $\frac{8}{24}$, or 1 to 3.
* *Remember, 12 inches = 1 foot.*

Find the following ratios in lowest terms. Write the answers as fractions.

3. 4 ft to 3 yd __________ 4. 1 gal to 3 qt __________

5. 2 hr to 40 min __________ 6. 5 oz to 1 lb __________

7. 1 cup to 12 oz __________ 8. 16 to 2 doz __________

Reprinted with special permission of King Features Syndicate, Inc.

Using Ratios

In some business situations, the ratio may be known but not the actual numbers involved in the comparison.

Example: The ratio of women workers to men workers is 4 to 9. If you know there are a total of 325 workers, then you can determine the number of women workers and the number of men workers. Since 4 out of 13 (4 + 9) workers are women and 9 out of 13 workers are men, $\frac{4}{13}$ of 325 will give the total number of women workers and $\frac{9}{13}$ of 325 will give the total number of men workers.

Multiply the fractional relationship times the total number of workers.

$$\frac{4}{13} \times \frac{325}{1} = 100 \text{ women workers}$$

$$\frac{6}{13} \times \frac{325}{1} = 225 \text{ men workers}$$

Find the following.

Use multioperations.

1. Divide a total of $25,000 into a 3:5 ratio. ________ ________

2. Divide 556 into a 3:1 ratio. ________ ________

3. Divide a total of 65 into a 2:3 ratio. ________ ________

4. Divide 393 into a 1:2 ratio. ________ ________

5. Divide a total of 225 into a 5:1 ratio. ________ ________

Proportion

A **proportion** is a statement that two ratios are equal, as in $\frac{1}{2} = \frac{2}{4}$. The statement can also be written 1:2 = 2:4 and is read: one is to two as two is to four.

Write each of the following as a proportion.

1. 14 compared to 28 is the same as 2 compared to 4.

____

2. 4 compared to 24 is the same as 7 compared to 42.

____

3. 2 compared to 5 is the same as 6 compared to 15.

____

4. 25 compared to 10 is the same as 10 compared to 4.

____

5. 350 compared to 70 is the same as 250 compared to 50.

____

One factor in a proportion may be missing. The missing factor can be found by using cross multiplication because the cross products are equal. A cross product in a proportion is found by multiplying the numerator of one ratio times the denominator of the other ratio.

$$\frac{2}{3} = \frac{4}{6}$$
$$2 \times 6 = 3 \times 4$$
$$12 = 12$$

To find a missing factor, follow these steps.

Example: $\frac{2}{3} = \frac{4}{?}$

Refer to page 1 for the definition of a number sentence.

Step 1: Cross multiply to form a number sentence.

$2 \times ? = 3 \times 4$

Step 2: Multiply to find the number each side will equal.

$3 \times 4 = 12$
$2 \times ? = 12$

Remember, division undoes what multiplication does.

Step 3: Divide to find the missing factor.

$2 \times ? = 12$
$? = 12 \div 2$
$? = 6$

Example: Five pens cost \$.90. What would nine pens cost? Think, 5 pens is to \$.90 as 9 pens is to ?

$\frac{5}{\$.90} = \frac{9}{?}$
$5 \times ? = 9 \times \$.90$
$5 \times ? = \$8.10$
$? = \$8.10 \div 5$
$? = \$1.62$

Find each missing factor.

6. $\frac{3}{17} = \frac{6}{?}$ ____________

7. $\frac{?}{20} = \frac{4}{80}$ ____________

8. $\frac{7}{12} = \frac{?}{96}$ ____________

9. $\frac{7}{?} = \frac{28}{25}$ ____________

10. $\frac{32}{4} = \frac{?}{8}$ ____________

EVALUATING YOUR SKILLS

The following problems evaluate your skills in the material covered in Units 2 through 16. Some of the problems will evaluate your ability to work problems mentally while others will evaluate your ability to use your calculator. Work the problems as quickly as you can.

1. 23,720.14 – 8,612.96 = ? ____________

2. Round $2.1447 to the nearest cent. ____________

3. Change $\frac{27}{35}$ to a percent. Carry out to the nearest tenth of a percent. ____________

4. Use fractions to multiply $7\frac{1}{2} \times 4\frac{2}{3}$. ____________

5. 23,671 ÷ 219.1 = ? __________

6. 400 is 20% less than ? __________

7. 6 is 1% of ? __________

8. Allocate $25,320 to the following departments:

   Dept. A: 45% __________

   Dept. B: 20% __________

   Dept. C: 35% __________

9. 453 units at $16 per C will cost how much? __________

10. 30 fluid ounces equals how many milliliters? __________

11. 6,221 × .01 = ? __________

12. Sam earns $160 a week plus 5% commission on all sales over $2,000. Sam sold $4,320 in one week. What were his gross earnings? __________

13. Use the sum-of-the years digits method to calculate the third year depreciation of a piece of equipment. The original cost was $6,300. Salvage value after four years is estimated to be $1,200. __________

14. Mike Hogshooter purchased a CB radio for $239. He paid 20% down. What is the amount to be financed? __________

15. The trade discount on an item listing for $1,239 is 10-20-5. Find the net price. __________

APPLYING YOUR KNOWLEDGE

Solve the following problems using the information given.

1. The Millers sold their home for \$56,000. They deducted a real estate broker's fee of \$2,240 to obtain their profit.
   a. Express as a ratio the broker's fee to the selling price.

   b. Change the ratio to a fraction.

   c. Change the ratio to a decimal.

   d. Change the ratio to a percent.

2. Two business partners divide their profits in a $\frac{3}{5}$ ratio. Find each partner's share of a \$10,400 profit.
   a. Partner A

   b. Partner B

3. Sam saves a ratio of \$1 to \$5 of everything he earns. If he earned a total of \$125 last week, how much did he put into his savings?

4. A building casts a shadow of 42 feet. A nearby mailbox pole 5 feet high casts a shadow of 7 feet. Find the height of the building.

5. The scale on a map is 3 inches equals 288 miles. What actual distance will $4\frac{1}{2}$ inches represent?

6. Ina earned $62 in four days. At that rate, how many days will it take her to earn $108.50?

____

7. Five packages of floppy disks cost $30. What will nine packages of floppy disks cost?

____

8. An advertising poster measuring $4\frac{1}{2}$ inches by 6 inches is to be enlarged. If the height of the enlargement is to be 10 inches, how wide will the new poster be?

____

9. Twenty-five feet of fencing wire weighs 14 pounds. Find the weight of 80 feet of the same kind of fencing wire.

____

10. Six boxes of envelopes cost $5.28. How many boxes can be purchased at the same rate for $11.44?

____

ADVANCED APPLICATIONS

Using the information given, calculate the correct answers.

1. Joan, Tom, Linda, and Bill received an inheritance of $20,000. They are to divide it by a ratio of 7:6:4:3. What is each person's share?

a. Joan ______

b. Tom ______

c. Linda ______

d. Bill ______

2. Adkins, Watkins, and Martinez go into business together investing $6,000, $7,000, and $12,000 respectively. The net profit for the year was $45,000. What amount of the profits should each person receive if the money is shared in proportion to each person's investment?

a. Adkins ______

b. Watkins ______

c. Martinez ______

3. Mr. Vickrum pays $395.75 taxes on his house valued at $55,750. Ms. Winston lives in a house valued at $64,200. If the tax rate is the same, what is the amount of tax on Winston's house?

______

4. The ratios of the corresponding sides of the two triangles are equal. Find the missing sides: A and B.

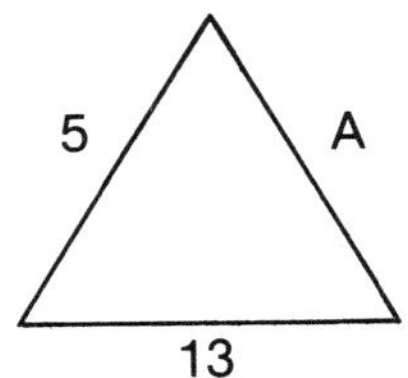

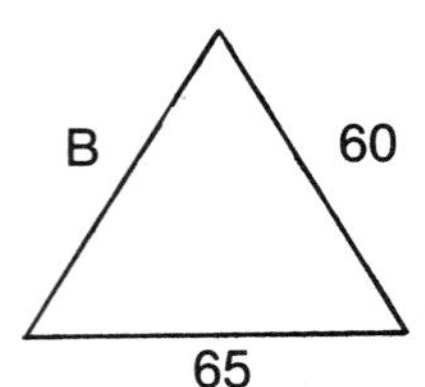

SIDE A: ____________________

SIDE B: ____________________

5. A wall hanging requires 1.2 meters of fabric. How many wall hangings can be made out of 3 yards of fabric? * *Answer in meters.*

3 yards × .9 = 2.7 meters ____________________

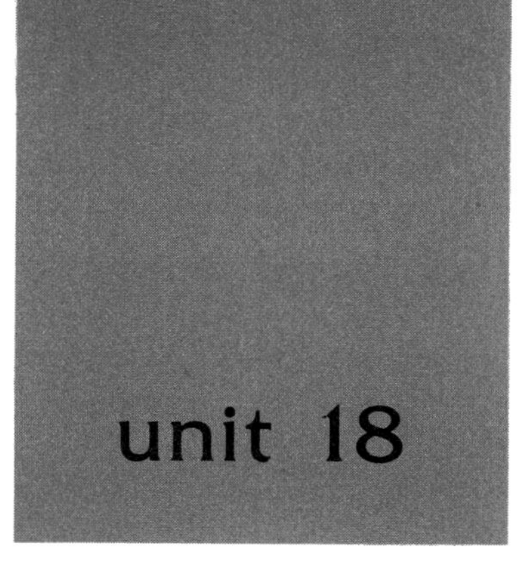

unit 18 Investments

"Anyone who stops learning is old—whether this happens at 8 or 80."

People invest their money primarily for financial security. Investments are usually safe if the investor is guaranteed to receive a certain rate of return. Certificate of deposits, savings accounts, and bonds are all relatively safe investments. Other types of investments offer the opportunity to make more money, but there is a greater risk involved. Many types of stocks, real estate, oil, and gas are risk investments. It is wise to compare investments to determine the amount of risk and how much you can afford to lose.

After studying this unit, you should be able to:

1. compute simple interest for various time periods
2. compute compound interest
3. use compound interest tables to compute interest
4. compute the total cost of the purchase or sale of stock
5. compute profit or loss from the sale of a bond

MATH TERMS AND CONCEPTS

The following terminology is used in this unit. Become familiar with the meaning of each term so that you understand its usage as you develop math skills.

1. **Bond**—a certificate issued by corporations and governmental agencies that promises to pay the face value at maturity plus periodic interest.
2. **Broker**—a person who handles the buying and selling of stocks, bonds, and/or real estate.
3. **Certificate of Deposit**—a savings account that yields a higher rate of interest because money is left in the account for a predetermined period of time.
4. **Commission**—a fee paid to a broker for services in buying and selling stocks, bonds, and/or real estate.
5. **Common Stock**—stock that has no stated dividend rate; dividends are paid to stockholders after preferred stockholders are paid.
6. **Compound Interest**—interest that is calculated on principal plus previous interest earned.

7. **Corporation**—a business formed as a separate legal entity from its owners who are known as stockholders.
8. **Dividend**—the part of the profits of a corporation that each holder of stock receives.
9. **Face Value**—the amount printed on a bond reflecting its worth at maturity.
10. **Interest**—the dollar amount earned over a specified period of time in an interest-bearing account.
11. **Interest-bearing Account**—account in which deposits earn interest.
12. **Investment**—savings that are put to work to earn income.
13. **Preferred Stock**—stock that earns dividends before dividends are paid on common stock.
14. **Principal**—the amount of an investment.
15. **Proceeds**—the selling price of stock minus the commission.
16. **Quarterly**—four times a year.
17. **Quoted Price**—the price at which a bond may be bought or sold.
18. **Regular Savings Account**—a savings account that earns a low interest rate and allows withdrawals at any time. Also called *passbook savings account* or *statement savings*.
19. **Semiannual**—twice a year.
20. **Stock**—a certificate issued by a corporation that represents ownership in the corporation.
21. **Stockholders (shareholders)**—people who own shares of stock in a corporation.

Interest-Bearing Savings Accounts

Banks offer various savings plans. A **regular savings account** may require a minimum amount, such as $200, to open and maintain the account. In some cases, these accounts may be called *passbook savings* or *statement savings* accounts. Money in a passbook account can be withdrawn at any time, usually without penalty. A **certificate of deposit** (CD) should also be considered when depositing money, especially if a higher rate of interest is desired. CDs require a deposit for a period of time during which the saver cannot withdraw money from the account without penalty.

All types of savings plans pay **interest**. The interest can be computed annually, semiannually, quarterly, or daily. Many savings plans offer **compound interest**. That is, interest is computed on the amount deposited by the customer as well as on any interest previously earned.

Simple Interest

Simple interest is paid only on the **principal** (the amount deposited) and not on any interest already earned.

Do you remember the formula for calculating simple interest given in Unit 16? Four elements are involved:

$$\text{Interest} = \text{Principal} \times \text{Rate} \times \text{Time}$$
$$I = P \times R \times T$$

Remember, the interest is the percentage, the principal is the base, and the rate is the percent. The fourth element is the time for which interest will be paid on the principal. The time is based on one year. Time of less than a year is stated as a fraction of a year. You are finding the interest for one year and dividing by the number of times the interest will be paid a year.

Note, the time is a fraction of a year.

Example: Interest computed **quarterly** is expressed as $P \times R \times \frac{1}{4}$. The first quarterly interest payment on $600 principal at 8% a year is:

$\$600 \times .08 \times \frac{1}{4} = \$12$

In the above example, if interest is paid *semiannually*, the **semiannual** payment is $\$600 \times .08 \times \frac{1}{2} = \$24$.

The new balance of an account is found by adding the interest to the principal. * ***Set decimal selector on 3 or 4 and round final answer to two places.***

Principal	+	Interest	=	New Balance
$600	+	$24	=	$624

Find the annual, quarterly, or semiannual interest earned for each of the following accounts.

	Deposit	Interest Rate	Compounded	Interest
1.	$826.17	$9\frac{3}{4}$%	semiannually	________
2.	$200	8%	annually	________
3.	$12,220.78	11%	quarterly	________
4.	$18,600	$8\frac{1}{2}$%	semiannually	________
5.	$1,725	$5\frac{3}{4}$%	annually	________
6.	$6,320.82	$10\frac{3}{4}$%	quarterly	________

Compound Interest

Compound interest is computed the same way as simple interest except that the interest is added to the principal for each time period. The interest for the second and succeeding time periods is calculated on both the interest and the principal.

To see how compound interest increases the balance of a savings account, study the following account.

Example: On July 1, a savings account is opened with $5,000. The interest rate is $8\frac{1}{4}\%$ compounded quarterly. Notice how the interest increases with each quarter.

	Amount Deposited	Interest	Balance
July 1	$5,000		$5,000.00
October 1		$103.13	$5,103.13
January 1		$105.25	$5,208.38
April 1		$107.42	$5,315.80

The chart above is calculated as follows. * ***Decimals are rounded to two places.***

October 1 $\$5,000 \times .0825 \times \frac{1}{4} = \$103.13$
$\$5,000 + \$103.13 = \$5,103.13$

January 1 $\$5,103.13 \times .0825 \times \frac{1}{4} = \$105.25$
$\$5,103.13 + \$105.25 = \$5,208.38$

If simple interest were paid for the same time, the balance would be $5,309.38.

April 1 $\$5,208.38 \times .0825 \times \frac{1}{4} = \$107.42$
$\$5,208.38 + \$107.42 = \$5,315.80$

Find the account balance after one year for each of the following problems.

	Deposit	Interest Rate	Compounded	Balance at End of One Year
1.	$2,000	7%	annually	____________
2.	$2,000	$5\frac{3}{4}\%$	quarterly	____________
3.	$2,000	$5\frac{3}{4}\%$	semiannually	____________

Calculator Procedure

Use the percent add-on feature on your calculator and the multi-operation procedure in computing compound interest. Find the rate of interest for a single payment period by dividing the yearly rate by the number of payments. * ***Carry the rate out to four places.***

Example: Assume $500 is invested at $8\frac{1}{4}\%$ interest. Interest is compounded quarterly. What is the balance at the end of the first year?

$8\frac{1}{4}\% \div 4 = 2.0625\%$ (single payment rate)

1. Set the decimal selector on 2.
2. Enter $500 and operate the multiplication key.
3. Enter 2.0625 and use the percent add-on feature. Did you get $10.31 interest and $510.31 new balance?
4. Operate the multiplication key and repeat Step 3 each time the interest is compounded. Did you get $542.54 for the balance at the end of the first year?

Solve the following problems using the percent add-on feature on your calculator.

	Deposit	Interest Rate	Compounded	Time Period	Balance
1.	$1,000	8%	semiannually	3 years	________
2.	$7,000	$7\frac{1}{2}\%$	quarterly	2 years	________
3.	$15,000	$8\frac{1}{2}\%$	quarterly	3 years	________
4.	$9,000	$5\frac{3}{4}\%$	semiannually	4 years	________
5.	$12,000	12%	annually	10 years	________

Compound Interest Tables

Computing compound interest is complicated and lengthy. To shorten the process, tables are used for interest compounded annually, semi-annually, and quarterly. Computations for compounding daily are so lengthy that only the use of computers makes daily compounding feasible.

The compound interest table, Illustration 18-1, shows the value of $1 if interest is compounded for a given number of time periods. Find the number of time periods involved and multiply that value times the amount in the account.

	$5\frac{1}{4}\%$		$5\frac{3}{4}\%$
1	1.013 125 0000	1	1.014 375 0000
2	1.026 422 2656	2	1.028 956 6406
3	1.039 894 0579	3	1.043 747 6923
4	1.053 542 6674	4	1.058 751 7683
5	1.067 370 4149	5	1.073 971 3250
6	1.081 379 6516	6	1.089 409 6628
7	1.095 572 7595	7	1.105 069 9267
8	1.109 952 1520	8	1.120 955 3068
9	1.124 520 2740	9	1.137 069 0394
10	1.139 279 6026	10	1.153 414 4068
11	1.154 232 6473	11	1.169 934 7389
12	1.169 381 9508	12	1.186 813 4133
13	1.184 730 0889	13	1.203 873 8561
14	1.200 279 6714	14	1.221 179 5428
15	1.216 033 3420	15	1.238 733 9987
16	1.231 993 7797	16	1.256 540 8000
17	1.243 163 6980	17	1.274 603 5740
18	1.264 545 8466	18	1.292 926 0003
19	1.281 143 0108	19	1.311 511 8116
20	1.297 958 0128	20	1.330 364 7939
21	1.314 993 7117	21	1.349 488 7878
22	1.332 253 0042	22	1.368 887 6891
23	1.349 738 8249	23	1.388 565 4496
24	1.367 454 1470	24	1.408 526 0780
25	1.385 401 9826	25	1.428 773 6404
26	1.403 585 3837	26	1.449 312 2614
27	1.422 007 4418	27	1.470 146 1252
28	1.440 671 2895	28	1.491 279 4757
29	1.459 580 1002	29	1.512 716 6182
30	1.478 737 0890	30	1.534 461 9196
31	1.498 145 5133	31	1.556 519 8097
32	1.517 808 6731	32	1.578 894 7819
33	1.537 729 9120	33	1.601 591 3944
34	1.557 912 6171	34	1.624 614 2707
35	1.578 360 2202	35	1.647 968 1009
36	1.599 076 1980	36	1.671 657 6423
37	1.620 064 0731	37	1.695 687 7209
38	1.641 327 4141	38	1.720 063 2319
39	1.662 869 8364	39	1.744 789 1409
40	1.684 695 0030	40	1.769 870 4848

Illustration 18-1. Compound Interest Table.

Example: $5,000 is deposited for five years with interest compounded quarterly at $5\frac{3}{4}$%. At the end of five years, interest will have been paid 20 times ($5 \times 4 = 20$). Read down the $5\frac{3}{4}$% chart to 20 periods and find 1.330 364 7939. This is multiplied by $5,000 to obtain an ending balance of $6,651.82 ($5,000 principal + $1,651.82 interest).

Use the compound interest table (Illustration 18-1) to solve the following problems.

	Deposit	Rate	Compounded	Balance End of	
1.	$7,000	$5\frac{3}{4}$%	quarterly	1 year	____________
2.	$2,000	$5\frac{1}{4}$%	quarterly	3 years	____________
3.	$20,000	$5\frac{3}{4}$%	quarterly	10 years	____________
4.	$8,175	$5\frac{1}{4}$%	quarterly	6 years	____________
5.	$8,175	$5\frac{3}{4}$%	quarterly	6 years	____________

Stock

People who buy or sell **stock** generally use a brokerage firm. **Brokers** charge a **commission** for stocks bought and sold (traded).

Example: Most newspapers print a list of the major stocks traded. The price of ACX stock is quoted at 17¾ ($17.75) per share in the newspaper. If you purchase 100 shares, the cost will be $1,775 ($17.75 × 100). Usually a commission is paid on each trade and added to the cost of the purchase.

Buying Stock

The total cost of a stock purchase depends on the cost per share, the number of shares purchased, and the broker's commission.

Number of Shares × Cost per Share + Commission = Total Cost

Example: Ten shares of Zinc. Inc. were purchased for $14\frac{1}{4}$ each. The commission was $22.50. Total cost is:

$10 \times \$14.25 + \$22.50 = \$165.00$

Depending on the brokerage firm, the commission may be a flat rate, a percent of total sales, or a combination.

Find the total cost for each of the following stock purchases.

	Number of Shares	Quoted Price per Share	Commission	Total Cost
1.	60	15½	2%	____________
2.	45	11¾	$12.50	____________
3.	20	48	$3\frac{1}{2}$%	____________
4.	150	50¼	4%	____________
5.	90	25⅜	$20.00	____________

Selling Stock

The sale of stock may result in either a profit or a loss. When stock is sold for more than what was paid, minus the commission, the result is a profit. When stock is sold for less than what was paid, minus the commission, the result is a loss.

Example: If 500 shares of stock were purchased at the quoted price of 16¼ plus a commission of $52 and the stocks were sold at the quoted price of 16¾ minus a commission of $64, the profit is figured as follows:

Total cost: 500 × $16.25 + $52 = $8,177.00
Total **proceeds**: 500 × $16.75 − $64 = $8,311.00
Profit on the sale: $8,311 − $8,177 = $134.00

Find the total proceeds and profit or loss on each of the following. Indicate a loss with parentheses.

	Total Cost	Number of Shares	Quoted Sale Price per Share	Commission Paid	Proceeds	Profit or Loss
1.	$1,975.40	70	55½	$80.00	____________	____________
2.	$2,723.80	120	22⅜	$56.39	____________	____________
3.	$2,267.43	150	16	$52.80	____________	____________
4.	$5,583.68	60	98¼	$129.69	____________	____________
5.	$7,322.58	200	33⅛	$146.41	____________	____________

When a **bond** is purchased, the buyer is lending money to a corporation or to the government. Interest is paid periodically for the time the bond is held, usually annually or semiannually.

The bond is purchased for a certain length of time, usually 10 to 30 years. At the end of the time period, the owner of the bond will be paid back the **face value** of the bond.

The **quoted price** for a bond may be more than, equal to, or less than the **face value**. The price of a bond is quoted as a percent of its face value.

$$\text{Percent} \times \text{Face Value} = \text{Bond Cost}$$

Example: A $1,000 bond is quoted at 82¾. The price of the bond is

$\$1{,}000 \times 82.75\% = \$827.50$

The annual interest for a bond is the face value times the interest rate:

$$\text{Face Value} \times \text{Interest Rate} = \text{Annual Interest}$$

Example: If the interest on a $1,000 bond is 9%, what is the annual interest?

$\$1{,}000 \times 9\% = \$90$

Example: If the interest on a $1,000 bond is 9%, what is the semiannual interest?

$9\% \div 2 = 4.5\%$
$\$1{,}000 \times 4.5\% = \$45$

Find the cost and interest payment for each of the following bonds.

	Face Value	Quoted Price	Cost	Interest Rate	Interest Payment
1.	$2,000	73¾	________	$7\frac{1}{2}\%$	________
2.	$15,000	89	________	9%	________
3.	$20,000	94⅖	________	$12\frac{3}{4}\%$	________
4.	$10,000	83½	________	$11\frac{5}{8}\%$	________
5.	$5,000	102	________	14%	________ semiannually

EVALUATING YOUR SKILLS

The following problems evaluate your skills in the material covered in Units 1 through 17. Some of the problems will evaluate your ability to work problems mentally while others will evaluate your ability to use your calculator. Work the problems as quickly as you can.

1. $40 \times ? \times 30$ __________

2. Use fraction form to divide $\frac{2}{7}$ by $\frac{7}{14}$. __________

3. Convert $5\frac{1}{2}$ feet to meters. __________

4. Multiply: $2.48 \times .10$ __________

5. Change $\frac{4}{9}$ to a decimal to four places. __________

6. 30 is 20% more than ? __________

7. What is the cost of 12,500 lb of fertilizer at $92 per T? __________

8. Carla Murphy is paid a weekly salary of $9.25 an hour on the basis of a 40-hour week with time and a half for overtime. During one week she works 47 hours. What are her total gross earnings? __________

9. Two partners share the profits in a ratio of 8 to 7. How much will each partner receive after dividing a profit of $14,500? __________ __________

10. What is the average of Jim's first four months of sales if sales were $5,320; $4,622; $6,379; and $5,983? __________

11. A loan of $1,200 is repaid in 24 equal installments of $65 each. What was the finance charge? ______

12. A retailer purchased an item for $120 with a markup of 43% on cost. What is the selling price of the item? ______

13. The Radio Shoppe lists speakers at $720 less a trade discount of 20-10-5. What is the invoice price? ______

14. Sandra's bank statement shows a balance of $329.15. Her checkbook balance is $293.20. A service charge of $4.50 is shown on the bank statement. There is one outstanding check for $40.45. What is the reconciliation balance? ______

15. An asset costing $9,000 has an estimated life of four years and a salvage value of $850. If depreciation is calculated using the straight-line method, what is the annual depreciation? ______

APPLYING YOUR KNOWLEDGE

Solve the following problems. Read each one carefully.

1. The Fashionable Shop has $8,311 in a savings account that pays $5\frac{3}{4}$% simple interest. How much interest will the account earn in one year?

   ______

2. The RGH Company has $15,650 in a certificate of deposit that pays $10\frac{1}{2}$% simple interest. How much interest will be earned in one quarter?

   ______

3. John Lester put $10,000 in a certificate of deposit that pays 12% simple interest semiannually. How much interest will he earn in six months?

   ______

4. Norma Medina deposited $2,679 on January 1 in a savings account that pays $5\frac{1}{4}$% interest compounded annually. How much interest will she earn at the end of three years?

____

5. On January 2, Ping-lin Ho made a deposit of $4,000 in a savings account. The interest is $8\frac{1}{2}$% per year compounded quarterly. Find the balance of his account on July 1 if interest was compounded on April 1 and July 1.

____

6. Citizens National Bank pays 7% interest compounded quarterly. Jean Hollenbeck deposited $630 on July 1. On October 1, she withdrew $150. On January 2, she deposited $300. What is her balance at the end of one year?

____

7. The Lowenstern Factory Gift Shop owns a $27,950 certificate of deposit that pays $11\frac{3}{4}$% interest. How much interest will the account earn at the end of one year if interest is:
   a. computed as simple interest

____

   b. compounded quarterly

____

8. Shannon Fry purchased 150 shares of stock quoted at 72¾. He sold them at 81½ a year later. During that time he was paid a dividend of $1.25 per share.

a. How much did he pay for the stock? ______

b. How much did he earn in dividends? * *Number of Shares × Dividend per Share = Total Dividend.* ______

c. How much was the total sale of the stock if the commission was $52? ______

d. What was the total profit? ______

9. Ouy Lin, Inc., a Japanese firm, issued fifty $1,000 bonds that pay $9\frac{3}{4}$% interest payable April 1 and October 1. What is the semi-annual interest payment? ______

10. At the C & W Brokerage firm, the commission charged on the purchase or sale of stock is based on the amount of the transaction. The table below is used to calculate the commission.

Amount of Transaction	Commission
Under $100	as mutually agreed
$100 to $500	2% + $5.00
$500 to $1,000	$1\frac{1}{2}$% + $10.00
$1,000 to $2,000	$1\frac{1}{10}$% + $15.00
$2,000 to $3,500	$\frac{1}{2}$% + $20.00
$3,500 and above	$\frac{1}{10}$% + $30.00

Find the commission and proceeds for the following transactions.

	Number of Shares	Quoted Price	Selling Price	Commission	Proceeds
a.	100	17½	______	______	______
b.	300	32⅝	______	______	______
c.	90	53¾	______	______	______

ADVANCED APPLICATIONS

Solve the following problems using the information given.

1. The Young Toy Store opened a savings account with a deposit of $8,350 on July 5. The bank pays $5\frac{1}{4}$% interest compounded daily if the account balance does not go below $500. The bank computes the interest after transactions are completed for the day. The interest is paid monthly. Using the information below, determine the amount of interest earned and the balance on the date indicated.

   * *Hint: There are 365 days in a year. Set decimal selector on 5 or FL for the rate of interest.*

Date	Transaction	Balance	Interest Earned	Balance
July 5				$8,350.00
7			________	________
8	Deposited $178	________	________	________
9	Withdrew $1,245	________	________	________
11			________	________
12	Deposited $3,687	________	________	________

2. Nelson plans to purchase $1,000 bonds quoted at 97½ plus $5.50 commission per bond. The interest is $10\frac{1}{2}$%.

   a. Nelson wants an annual income of a least $800 from the investment. How many bonds will he have to buy?

   ____

   b. What is Nelson's total investment?

   ____

3. Mary Mancellous purchased 150 shares of stock quoted at 22¼ plus $36.69 for commission. During the year she received four quarterly dividends of $.22 per share.

a. How much did she invest in the stock? ________________

b. How much did she receive in dividends? ________________

c. She sold the stock for $17.50 per share minus $33.13 commission. What was the net loss on the sale? ________________

The Hodgepodge Shop

series IV A Retail Application

"If you don't stand for something you are likely to fall for anything."

Welcome back to The Hodgepodge Shop. You will continue your training by working in the accounting department.

After completing this training program, you should be able to:

1. compute finance charges
2. compare credit terms
3. complete depreciation schedules

TERMINOLOGY

The following terminology is used in this application. Become familiar with the meaning of this term so that you understand its usage.

Amortization Schedule—a table showing the periodic payments and decreasing balance of a loan.

ON-THE-JOB-TRAINING

In this segment of your training you will help The Hodgepodge Shop decide how to finance and depreciate the purchase of five new computers, printers, and software for the main store and branch stores. Read each job assignment carefully. Follow each step outlined for you. Do not skip any steps.

Good luck. Remember, you are a trainee so ask for help along the way as you need it.

Assignment 1: Credit Plans

Management is considering various ways to finance the five computers. Each computer system costs $9,671. The Hodgepodge Shop has the choice of financing the computers through Computer's Inc., the place of purchase; Suburban Bank and Trust; or Federal Farm Union Bank.

April 15

1. Remove the worksheet (Form 1, page 327). Compute the total cost of the computers and record on line 1.
2. The Hodgepodge Shop plans to make a $10,000 down payment. Record the amount of the down payment on line 2 and the amount to be financed on line 3.
3. Computer's Inc. will finance the computers for three years at a stated rate of 12%. Compute the total finance charge and record on line 4. Compute the total amount to be repaid and record on line 5. Find the monthly payment and record on line 6.
4. As a comparison with other sources of financing, The Hodgepodge Shop asks you to determine the annual percentage rate. When charts are not available, the following formula is used to find the approximate rate:

$$\frac{2 \times \text{Number of Payments in One Year} \times \text{Finance Charge}}{\text{Principal} \times (\text{Number of Payments} + 1)}$$

   Use the formula and record the annual percentage rate on line 7.
5. Suburban Bank and Trust will finance the computers for two years at 16.5% APR. Suburban Bank and Trust has sent a payment schedule showing the amount of each monthly payment applied to the principal and interest. (See Form 2.) Record the amount of the monthly payment on line 8. Find the total amount of interest to be paid and record on line 9.
6. Federal Farm Union Bank will finance the computers for three years at 12% APR. Federal Farm Union Bank has also sent a payment schedule. (See Form 2.) Record the amount of the monthly payment on line 10. Find the total amount of interest to be paid and record on line 11.
7. File Form 2.

Assignment 2: Depreciation

Management needs to decide which method of depreciation to use: straight-line, declining-balance, or sum-of-the-years-digits. The five computers have an estimated trade-in value at the end of four years of $1,250 each. Management will use depreciation schedules to decide which depreciation method to use. You have been asked to complete the depreciation schedules for the three methods.

April 15

1. Remove the Depreciation Schedules (Form 3). Complete the depreciation schedule for straight-line, declining-balance, and sum-of-the-years-digits. Depreciation is calculated using the cost of one computer.

2. Submit the worksheet (Form 1) and depreciation schedules (Form 3) to your supervisor.

JOB EVALUATION

The job evaluation form would usually be completed by your supervisor. However, you are given an opportunity to rate yourself. Look at your weaknesses and talk with your supervisor about how you can improve. This evaluation form is used as a basis for hiring full-time employees.

April 20

1. Remove the Performance Evaluation (Form 4).
2. Rate yourself on a scale from 1 to 10. Read each statement carefully and check the number column that you feel applies to you. Work hard to improve any weaknesses.
3. Submit Form 4 to your supervisor.

WORKSHEET

Form 1

1. Total Cost of Computers ____________

2. Amount of Down Payment ____________

3. Total Amount to be Financed ____________

Computer's, Inc.

4. Total Finance Charge ____________

5. Total Amount to be Repaid ____________

6. Monthly Payment ____________

7. Annual Percentage Rate ____________

Suburban Bank and Trust

8 Monthly Payment ____________

9 Total Interest ____________

Federal Farm Union Bank

10. Monthly Payment ____________

11. Total Interest ____________

SUBURBAN BANK AND TRUST—AMORTIZATION SCHEDULE

Loan Amt = $38,355.00, # of Periods = 24, Int. Rate = 16.5%

Pymt	Monthly	Principal	Interest	Balance
1	$1,887.16	$1,359.78	$527.38	$36,995.20
2	1,887.16	1,378.48	508.68	35,616.70
3	1,887.16	1,397.43	489.73	34,219.30
4	1,887.16	1,416.64	470.52	32,802.70
5	1,887.16	1,436.12	451.04	31,366.60
6	1,887.16	1,455.87	431.29	29,910.70
7	1,887.16	1,475.89	411.27	28,434.80
8	1,887.16	1,496.18	390.98	26,938.60
9	1,887.16	1,516.75	370.41	25,421.90
10	1,887.16	1,537.61	349.55	23,884.30
11	1,887.16	1,558.75	328.41	22,325.50
12	1,887.16	1,580.18	306.98	20,745.30
13	1,887.16	1,601.91	285.25	19,143.40
14	1,887.16	1,623.94	263.22	17,519.50
15	1,887.16	1,646.27	240.89	15,873.20
16	1,887.16	1,668.90	218.26	14,204.30
17	1,887.16	1,691.85	195.31	12,512.50
18	1,887.16	1,715.11	172.05	10,797.30
19	1,887.16	1,738.70	148.46	9,058.64
20	1,887.16	1,762.60	124.56	7,296.04
21	1,887.16	1,786.84	100.32	5,509.20
22	1,887.16	1,811.41	75.75	3,697.79
23	1,887.16	1,836.32	50.84	1,861.47
Last Payment Adjusted To		1,887.07		
24	1,887.07	1,861.47	25.60	0.00

FEDERAL FARM UNION BANK—AMORTIZATION SCHEDULE

Loan Amt = $38,355.00, # of Periods = 36, Int. Rate = 12%

Pymt	Monthly	Principal	Interest	Balance
1	$1,273.94	$ 890.39	$383.55	$37,464.60
2	1,273.94	899.29	374.65	36,565.30
3	1,273.94	908.29	365.65	35,657.00
4	1,273.94	917.37	356.57	34,739.70
5	1,273.94	926.54	347.40	33,813.10
6	1,273.94	935.81	338.13	32,877.30
7	1,273.94	945.17	328.77	31,932.10
8	1,273.94	954.62	319.32	30,977.50
9	1,273.94	964.16	309.78	30,013.40
10	1,273.94	973.81	300.13	29,039.60
11	1,273.94	983.54	290.40	28,056.00
12	1,273.94	993.38	280.56	27,062.60
13	1,273.94	1,003.31	270.63	26,059.30
14	1,273.94	1,013.35	260.59	25,046.00
15	1,273.94	1,023.48	250.46	24,022.50
16	1,273.94	1,033.72	240.22	22,988.80
17	1,273.94	1,044.05	229.89	21,944.70
18	1,273.94	1,054.49	219.45	20,890.20
19	1,273.94	1,065.04	208.90	19,825.20
20	1,273.94	1,075.69	198.25	18,749.50
21	1,273.94	1,086.44	187.50	17,663.10
22	1,273.94	1,097.31	176.63	16,565.80
23	1,273.94	1,108.28	165.66	15,457.50
24	1,273.94	1,119.37	154.57	14,338.10
25	1,273.94	1,130.56	143.38	13,207.50
26	1,273.94	1,141.86	132.08	12,065.70
27	1,273.94	1,153.28	120.66	10,912.40
28	1,273.94	1,164.82	109.12	9,747.58
29	1,273.94	1,176.46	97.48	8,571.12
30	1,273.94	1,188.23	85.71	7,382.89
31	1,273.94	1,200.11	73.83	6,182.78
32	1,273.94	1,212.11	61.83	4,970.67
33	1,273.94	1,224.23	49.71	3,746.44
34	1,273.94	1,236.48	37.46	2,509.96
35	1,273.94	1,248.84	25.10	1,261.12
Last Payment Adjusted To		1,273.73		
36	1,273.73	1,261.12	12.61	0.00

DEPRECIATION SCHEDULES
(Based on the cost of one computer)

Straight-Line

Annual Depreciation ____________

Year	Balance	Amount of Depreciation	Book Value	Accumulated Depreciation
1				
2				
3				
4				

Total Depreciation for the five computers ____________

Declining-Balance

Find the depreciation rate. * *Remember the useful life becomes the denominator and forms a fraction with one as the numerator.* ____________

Year	Balance	Amount of Depreciation	Book Value	Accumulated Depreciation
1				
2				
3				
4				

Total Depreciation for the five computers ____________

Sum-of-the-Years-Digits

Find the sum of the years. ______________

$4 + 3 + 2 + 1 = 10$

Year	Balance	Amount of Depreciation	Book Value	Accumulated Depreciation
1				
2				
3				
4				

Total Depreciation for the five computers ______________

Form 4

PERFORMANCE EVALUATION

THE HODGEPODGE SHOP

	Needs Improvement		Average –		Average		Average +		Excellent	
	1	2	3	4	5	6	7	8	9	10
PRODUCTION										
How accurate and complete work is										
	Careless. Often makes mistakes.		Some errors. Work usually passable.		Few errors. Average worker.		Good worker. Thorough.		Errors rare. Very thorough. Careful.	
How time is used										
	Loiters, or absent too much. Could accomplish much more.		Absent more than necessary. Could accomplish more.		Does standard amount of work.		Seldom absent. Steady worker. Standard amount.		Never absent. Regularly produces more than standard amount.	
ATTITUDE										
Toward others										
	Quarrelsome, jealous, uncooperative fault-finding.		Sometimes inconsiderate.		Normally gets along well.		Cooperative, friendly, helpful.		Well-liked by all. Tactful and considerate.	
Toward Hodgepodge										
	Constantly complains.		Puts self-interest first.		Takes some pride in company.		Supports company. Feels company supports employees.		Puts company first.	
Cooperation										
	Resents supervision.		Sometimes willing to accept criticism.		Complies with instructions.		Willing to consider suggestions.		Appreciates suggestions. Accepts criticism.	
DEPENDABILITY										
Reliability and attendance										
	Cannot be relied on. Often late.		Needs frequent reminders. Punctuality fair.		Dependable. Attendance good.		Can be counted on. Usually prompt.		Work done on time. Good attendance.	

appendix A General Calculator Instructions

There are many makes and models of electronic calculators on the market today. Some have a display so that the answer appears on a screen. Some have a printout tape. Others have both a display and a tape. Differences occur in the location of function keys, tape symbols, and special features.

Do not attempt to learn everything about your calculator at one time. Practice only the section of the operating manual that covers what you are studying in the text.

Read, *carefully*, the following pages each time you begin to use a different calculator. The following outline will help you get acquainted with a new calculator.

PREPARATORY STEPS

1. Clear your desk. You need only your book, calculator, and a pen or pencil.
2. If you use your right hand for the calculator, place your calculator at a slight angle on the right with this book positioned to the left of the calculator. Reverse positions if your use your left hand for the calculator.
3. Sit erect. Good posture is important in developing the necessary skill and speed to operate your calculator.
4. Keep your arm parallel to the slope of the machine. Curve your fingers over the home row: 4, 5, 6 keys.

TOUCH METHOD

Touch operation of the calculator requires using the same finger to operate a number without looking at the keys. Refer to Illustration A-1 for the correct fingering.

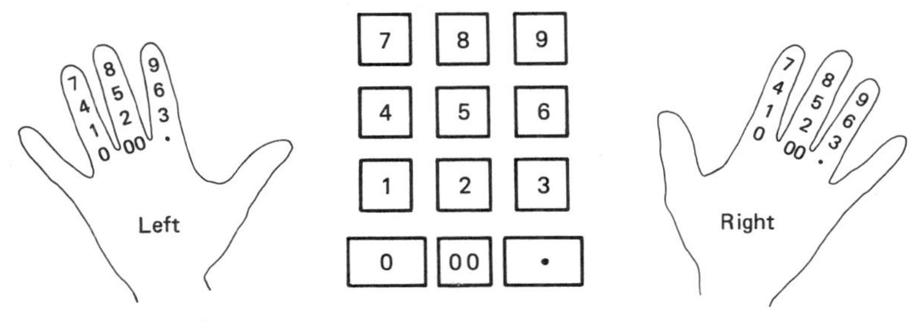

Illustration A-1.

The zero, double zero, and decimal point may be at a slightly different angle. Experiment to determine which finger is most comfortable.

1. Locate the *on/off switch.*
2. Locate the *Paper Advance Selector.* When operated, the paper advances.
3. Locate the *Decimal Selector.* Most calculators are equipped with a decimal selector that will round automatically to the number of decimal places set.
   a. The F or FL position refers to the floating decimal. Answers will not be rounded but will be carried out to the maximum places allowed by the calculator.
   b. Does your calculator have a *Plus* (+), AM, or A position or selector? This is the *Add Mode* used in working with dollars and cents. The calculator will automatically put in two decimal places for all addition and subtraction entries. A few models are equipped with an ×'s mode which will enter two decimals in the multiplier only. If a decimal appears in the multiplicand, it must be entered manually.
4. Locate the *Total Key.* This will be indicated by either Total, *, T, CLR, C, or Clear. (Do not confuse a separate clear or clear-entry key with the Total Key.)
5. Locate the *Subtotal Key.* Printing calculators will give a total or running balance without clearing the register. The key is indicated by the letter S or a diamond (◊). Display models show the balance on a screen after each entry and do not have a separate subtotal key.
6. Locate the *Clear Key.* Learn how to clear the last entry made *and* all registers of your calculator. Some calculators will have a clear key and a clear-entry (CE) key, while others combine the functions into one key. Operating a CE key once will clear the last entry made on the keyboard. Operating it twice will clear the entire register. A few models will also clear the memory with the CE key.
7. Locate the *Rounding Selector.* Many calculators will automatically round all answers according to the number of decimals set. Some models have a selector. The 9 or plus automatically rounds up. The arrow pointing down or a 0 cuts off at the number of decimals set. The 5/4 or 5 position rounds all answers following these rules:
   a. If the first number to the right of the number of decimals set is 5 or more, 1 is added.
   b. If the number to the right is less than 5, that number is dropped.
8. Locate the *Percent Key.* Percents do not have to be converted to decimals when you use this key. The percent key can be used to calculate the percentage, rate, and base. Use the percent key instead of the equals key to obtain the answer. Most models have the add-on and discount capabilities. To add-on the %, operate the plus key after operating the percent key. To obtain the net amount for a discount, operate the minus key after operating the percent key. * ***Clear the calculator: some models will reenter the amount of the percent when the total key is operated.***

9. Locate the *Constant Position*. Almost all calculators have a constant operation mode. While on some calculators it is programmed in, others have a dial or switch that must be set. To determine if your calculator uses first or second entry as a constant, follow this procedure:
   a. Set your calculator on the constant operation (if required).
   b. Set the decimal selector on 0.
   c. Using the example problem: $5 \times 4 = 20$
      $5 \times 2 = 10$
      (1) Enter 5.
      (2) Operate the multiplication key.
      (3) Enter 4 and operate the equals key. The product 20 will show on your tape or screen. *Do not clear your calculator.*
      (4) Enter 2 and operate the equals key. (Do not enter the multiplication key again.) If the product is 10, the first entry is the constant: $5 \times 2 = 10$. If the product is 8, the second entry is the constant: $4 \times 2 = 8$. The second entry is the constant in division. Follow the same procedure as constant multiplication.
10. Locate the *Non-Add Key*. This is found on printing calculators. It allows you to print numbers without performing any mathematical operations. Some models will also print the date. Check your operating manual for instructions on these special features.
11. Locate the *Memory Key* or *Grand Total Key*. These features allow you to store information.
   a. Memory Registers:
      (1) *Memory plus key*. Entries are added to the contents of memory.
      (2) *Memory minus key*. Entries are subtracted from the contents of memory.
      (3) *Memory subtotal key*. The contents of memory are printed or displayed without clearing the register.
      (4) *Memory total key*. Gives the total and clears the register.
      (5) Does your calculator have a *Sigma* (Σ) switch or dial? When set, all products and quotients are automatically accumulated in the memory register. To take a grand total, operate the memory total key.
   b. *Grand Total* or *Accumulator Selector*: Position your calculator on GT (usually a slide selector) to accumulate all sums and differences. Some models will also accumulate products and quotients. Learn the capabilities of your calculator.
   c. *Grand total* with *Equals/Plus* and *Equals/Minus Keys*: These will give the product or quotient to an individual problem and automatically add or subtract the answer to memory.
      Some models have more than one memory register. Learn the advantages of using each register.
   d. *Non-Memory Calculators*: Some calculators without memory will accumulate. After each multiplication or division problem, operate the plus key. Operate the total key to obtain the grand total.

SPECIAL FEATURES

Many calculators have special features. Some of the more common are:

1. **Average Key**. After the sum of a series of numbers is taken, operate the average key to compute the simple average.
2. **Item Count Key**. Counts the number of times the plus key is operated in addition. Each time the minus key is used, 1 is subtracted from the counter. Some models will have a switch to count the number of times the plus and minus keys are operated. The register is automatically cleared when the total key is operated.
3. **Delta Key**. Used to calculate the percent increase/decrease between two time periods. Enter the first time period, operate the delta key. Enter the second time period and operate the equals key to find the percent increase/decrease. Be sure to read your operating manual for specific instructions.
4. **Markup/Markdown (MU/MD)** or **Gross-Margin Key**. Used to calculate the cost or selling price when the percent of markup or markdown is based on the unknown quantity.

   Example: Cost of item is $65. Percent of markup is 40% based on the selling price.
   a. Enter 65 and operate the MU/MD or GM key.
   b. Enter 40 and operate the equals key.
   The amount of markup is $43.33 indicated by the delta sign (Δ).
   The selling price is $108.33.

   Example: Selling price is $300 and percent of markup is 40% based on the cost.
   a. Enter 300 and operate the MU/MD or GM key.
   b. Enter 40 and operate the +/− key and equals key *or* enter −40 and operate the equals key.
   The amount of markup is $85.71 indicated by the delta sign (Δ).
   The cost is $214.29.
   * ***This procedure is also used to calculate markdown.***

Now, take a look at your calculator. Are there keys, dials, or selectors that have not been discussed? Be sure to read the operating manual before asking your teacher for an explanation.

appendix B Metric Conversion Table

The following table can be used to convert from English to metric and from metric to English. The calculations should be made from left to right and are approximate because of the number of decimal places used.

When You Know	Multiply by	To Find
	Length and Distance	
English System		
inches	25	millimeters
inches	2.5	centimeters
feet	30.5	centimeters
feet	0.3	meters
yards	0.9	meters
miles	1.6	kilometers
Metric System		
millimeters	0.04	inches
centimeters	0.4	inches
centimeters	0.03	feet
meters	3.3	feet
meters	1.1	yards
kilometers	0.6	miles
	Area	
English System		
square inches	6.5	square centimeters
square feet	929	square centimeters
square feet	0.09	square meters
square yards	0.8	square meters
acres	0.4	hectares
square miles	2.6	square kilometers
Metric System		
square centimeters	0.2	square inches
square meters	10.8	square feet
square meters	1.2	square yards
hectares	2.5	acres
square kilometers	0.4	square miles

When You Know	Multiply by	To Find
	Mass and Weight	
English System		
ounces	28	grams
pounds	0.45	kilograms
short tons	0.9	metric tons
Metric System		
grams	0.035	ounces
kilograms	2.2	pounds
metric tons	1.1	short tons
	Volume and Capacity	
English System		
fluid ounces	30	milliliters
pints	0.47	liters
quarts	0.95	liters
dry quarts	1.1	liters
gallons	3.8	liters
Metric System		
milliliters	0.03	fluid ounces
liters	2.1	pints
liters	1.06	quarts
liters	0.9	dry quarts
liters	0.26	gallons

appendix C Decimal Equivalents

3rds	4ths	5ths	6ths	7ths	8ths	9ths	10ths	11ths	12ths	Equivalents
									1	.083333
								1		.090909
							1			.100000
						1				.111111
					1					.125000
				1						.142857
			1						2	.166667
								2		.181818
		1					2			.200000
						2				.222222
	1				2				3	.250000
								3		.272727
				2						.285715
							3			.300000
1			2			3			4	.333333
								4		.363636
					3					.375000
		2					4			.400000
									5	.416665
				3						.428572
						4				.444444
								5		.454546
	2		3		4		5		6	.500000
								6		.545455
						5				.555556
				4						.571429
									7	.583331
		3					6			.600000
					5					.625000
								7		.636364
2			4			6			8	.666667
							7			.700000
				5						.714287
								8		.727273
	3				6				9	.750000
						7				.777778
		4					8			.800000
								9		.818182
			5						10	.833333
				6						.857144
					7					.875000
						8				.888889
							9			.900000
								10		.909091
									11	.916663

appendix D Multiplication Table

1	2	3	4	5	6	7	8	9	10	11	12
2	4	6	8	10	12	14	16	18	20	22	24
3	6	9	12	15	18	21	24	27	30	33	36
4	8	12	16	20	24	28	32	36	40	44	48
5	10	15	20	25	30	35	40	45	50	55	60
6	12	18	24	30	36	42	48	54	60	66	72
7	14	21	28	35	42	49	56	63	70	77	84
8	16	24	32	40	48	56	64	72	80	88	96
9	18	27	36	45	54	63	72	81	90	99	108
10	20	30	40	50	60	70	80	90	100	110	120
11	22	33	44	55	66	77	88	99	110	121	132
12	24	36	48	60	72	84	96	108	120	132	144

Index